Healthy Oils

Healthy Oils

Fact versus Fiction

Myrna Chandler Goldstein
and Mark A. Goldstein, MD

GREENWOOD

AN IMPRINT OF ABC-CLIO, LLC
Santa Barbara, California • Denver, Colorado • Oxford, England

Copyright 2014 by ABC-CLIO, LLC

All rights reserved. No part of this publication may be reproduced, stored in a retrieval system, or transmitted, in any form or by any means, electronic, mechanical, photocopying, recording, or otherwise, except for the inclusion of brief quotations in a review, without prior permission in writing from the publisher.

Library of Congress Cataloging-in-Publication Data

Goldstein, Myrna Chandler, 1948–
  Healthy oils : fact versus fiction / Myrna Chandler Goldstein and Mark A. Goldstein, MD.
       pages cm
    Includes index.
    ISBN 978-1-4408-2875-1 (hardback) — ISBN 978-1-4408-2876-8 (ebook)
  1. Lipids in human nutrition.   2. Oils and fats, Edible—Health aspects.
  I. Goldstein, Mark A. (Mark Allan), 1947–   II. Title.
    QP751.G65   2014
    612.3'97—dc23   2014013447

ISBN: 978-1-4408-2875-1
EISBN: 978-1-4408-2876-8

18 17 16 15 14   1 2 3 4 5

This book is also available on the World Wide Web as an eBook.
Visit www.abc-clio.com for details.

Greenwood
An Imprint of ABC-CLIO, LLC

ABC-CLIO, LLC
130 Cremona Drive, P.O. Box 1911
Santa Barbara, California 93116-1911

This book is printed on acid-free paper ∞

Manufactured in the United States of America

We dedicate this book to our five grandchildren:
Aidan Zev Goldstein, born February 8, 2008
Payton Maeve Goldstein, born December 4, 2009
Milo Adlai Kamras, born August 9, 2011
Erin Abigail Goldstein, born December 29, 2011
Zoe "Scout" Eames Kamras, born February 16, 2014

Contents

Preface

Not that long ago, people were routinely told to reduce their consumption of all oils. No distinction was made between the healthier oils, such as olive oil and peanut oil, and the less than desirable oils, such as those found in vegetable oil. Today, we know that some oils actually support health. But, there are few, if any, books on the topic.

Having spent the past several years researching and writing *Healthy Foods: Fact versus Fiction* and *Healthy Herbs: Fact versus Fiction*, the next logical step was to address the dearth of books on oils. And, so we began to locate research on various oils. Eventually, we found sufficient studies on the following oils: almond, argan, avocado, borage seed, canola, coconut, cod liver, evening primrose, fish, flaxseed, jojoba, krill, olive, peanut, peppermint, pumpkin seed, red palm, rice bran, safflower, salmon, saw palmetto, sea buckthorn, sesame, walnut, and wheat germ. While here we have written only one entry on most of the oils, some, such as coconut oil and olive oil, had many more studies. So, they have multiple entries.

Although we have identified healthy properties in these oils, it is important to remember that all oils are high in caloric content. As a result, unless one is trying to gain weight, oils should be used in moderation. If the oils mentioned in this book are not readily available in your nearby stores, they may be easily purchased online.

Almond Oil

Almonds and almond oil may be traced to prebiblical times. In fact, almonds and almond oil are mentioned in the Bible. Even before the Christian era, they were believed to have medicinal properties. Today, almonds are grown primarily in India, Turkey, Israel, Syria, and in sections of eastern Mediterranean, northern Africa, and southern Europe. In the United States, they are grown in California. Almond oil is made from almond seeds.

Almonds and almond oil are believed to be useful for a wide variety of medical problems. They are thought to have anti-inflammatory, immunity-boosting, and antihepatotoxicity properties. They are said to support gastrointestinal and cardiovascular health, and they may be used to treat skin conditions such as psoriasis and eczema.[1]

Almond oil may be a little hard to find in traditional markets. However, it is sold in specialty stores and readily available online. Almond oil is moderately priced.

MAY HELP CONTROL WEIGHT AND REDUCE RISK FOR TYPE 2 DIABETES

In a study presented to the 2012 general meeting of the American Society for Microbiology, researchers from the Missouri University of Science and Technology described feeding oil derived from the seeds of wild almond trees to 14 normal and 14 obese laboratory mice. The researchers began by separating the mice into four groups. For nine weeks, one group of normal mice and one group of obese mice were fed a standard diet. During that same time, the diets of one group of normal mice and one group of obese mice were supplemented with 0.5% sterculic oil, extracted from the seeds of the wild almond tree. The researchers learned that adding sterculic oil to the diets of obese mice increased their glucose tolerance and insulin sensitivity. Apparently, this was the direct result of the oil's effect on three types of microorganisms that live in the gastrointestinal tracts of the mice. Meanwhile, the normal mice experienced no adverse effects.[2]

MAY HELP ADULTS WITH IMPAIRED GLUCOSE TOLERANCE (WHO ARE AT INCREASED RISK FOR TYPE 2 DIABETES)

In a randomized, five-arm, crossover study published in 2011 in *Nutrition & Metabolism*, researchers from Indiana (United States) wanted to determine if the

Almond oil and almond nuts with flowers.
(Margo555/Dreamstime.com)

consumption of almond products at breakfast would alter feelings of fullness and
stabilize blood glucose levels. The cohort consisted of 14 subjects who reported
to the laboratory five times, separated by at least one week. During the laboratory
visits, the subjects ate breakfasts that included orange juice, farina, and whole
almonds or almond butter or defatted almond flour or almond oil or no almond
products (control). After the test breakfasts were completed, blood was drawn
and appetite ratings were taken at 15, 45, 60, 90, 120, 180, and 240 minutes. At
240 minutes, the subjects ate lunches consisting of a plain white bagel, grape or
strawberry jam, and tap water. Following the completion of lunch, the researchers conducted several blood tests and appetite evaluations. The researchers found
that when compared to the control breakfast, the addition of whole almonds
to breakfast significantly increased satiety and decreased glucose concentrations
throughout the day. Almond oil had a similar, but lesser effect on postmeal blood
glucose concentrations. Both whole almonds and almond oil significantly reduced the insulin response after lunch.[3]

MAY SUPPORT LIVER HEALTH

In a study published in 2011 in *Food Chemistry*, researchers from China tested
the ability of almond oil to support liver health. They began by randomly dividing rats into several groups of eight. After setting aside one group of rats to
serve as the controls, the rats were treated with a liver protection medication
and different doses of almond oil before being treated with carbon tetrachloride,

a chemical that induces liver toxicity. The researchers found that the almond oil helped protect the liver from the carbon tetrachloride. "Pre-treatment with almond oil suppressed the acute hepatic damage and was consistent with an improvement in the serum biological parameters of hepatotoxicity."[4]

MAY HELP PROTECT SKIN

In a study published in 2007 in the *Journal of Cosmetic Dermatology*, researchers from New Delhi, India, wanted to learn if almond oil could protect the skin from ultraviolet-B radiation-induced chronic skin damage. The researchers began by dividing 80 adult Swiss female mice into four groups of 20. The mice in the first group served as the controls; they received no treatment. The mice in the second group received only topical almond oil treatment. The mice in the third group received only ultraviolet B radiation; and the mice in the fourth group were treated with topical almond oil and ultraviolet B radiation. During the first week, mice were exposed to 15 minutes of light per day. Each week, the exposure times were increased by 15 minutes. By the 12th week, the daily exposure time was capped at 3 hours. The researchers found that almond oil offered a good deal of protection to the mice that were exposed to the ultraviolet B radiation. "The results of this study illustrate that almond oil treatments helps to ameliorate and to partially inhibit some of the histologic damage associated with photoaging of skin and appears to contribute to a decrease in the prevalence of UVB-induced skin tumors in mice."[5]

MAY BE USEFUL FOR PRETERM INFANTS

In a randomized, controlled trial published in 2008 in *Child: Care, Health and Development*, researchers from France examined the use of topical almond and other oils for massaging low-risk preterm infants. The cohort consisted of 49 infants born between 31 and 34 weeks of gestation. Each infant was assigned to receive either no oil (control) or sweet almond oil, "used traditionally in the practice of oil massage for newborn infants in France," or a blend of sunflower, high-oleic sunflower, rapeseed, and grapeseed oils. All the infants received 15-minute massages twice each day for 10 consecutive days. The researchers found that the infants in all three of the treatment groups had better outcomes than the infants in the control group. "Massage with oil enhanced neurodevelopmental performance, improved the moisturization of the skin and reduced days in the hospital." At the same time, "no adverse dermatological events were observed."[6]

MAY BE USEFUL FOR PESTICIDE POISONING

In a study published in 2012 in *Human and Experimental Toxicology*, researchers from Iran noted that aluminum phosphide, a pesticide, has "become one of the commonest causes of poisoning and even suicide in developing countries including Iran and India." Could sweet almond oil serve as an anecdote to this deadly pesticide? The researchers decided to test their theory in rats. They began by placing 35 rats into four groups. The rats in the first group served as the controls; the rats in the three other groups received aluminum phosphide alone or with sweet almond oil with varying time intervals. The rats in the control group all survived, but they received no pesticide. Of the rats exposed to pesticide, the ones that were treated immediately after exposure with almond oil had the best outcomes. "The poisoned animals undergoing immediate intervention with sweet almond oil had better outcomes considering both mean survival times and complete survival rate."[7]

MAY HELP PREVENT STRIAE GRAVIDARUM IN WOMEN HAVING THEIR FIRST CHILD

In a study published in 2012 in the *Journal of Clinical Nursing*, researchers from Turkey wondered if bitter almond oil, with or without massage, could play a role in preventing striae gravidarum in women having their first child. (Striae gravidarum are the "stretch marks" so often seen in women who are pregnant.) The cohort consisted of 141 women; they were assigned to one of three groups. The first group applied bitter almond oil with massage to the abdominal region, thighs, and breasts; the second group applied bitter almond oil without massage; and the third group consisted of controls, who did not do anything to prevent the formation of striae gravidarum. The researchers recommended that the women in the massage group massage the oil for 15 minutes every other day from the 19th week of pregnancy until the 32nd week. After the 32nd week, they were asked to massage the oil every day, preferably in the evening. The women in the oil only group followed a similar schedule; only they applied the oil without any massage. The women who applied the oil with massage had the lowest amounts of striae gravidarum in the abdominal area; the women who only applied the oil had less than the control women, but the difference was relatively small. "Pregnant women could learn from the findings of this study. Nurses and midwives could also benefit from these findings in their practices."[8]

IS ALMOND OIL BENEFICIAL?

Almond oil appears to have a number of beneficial properties. However, it is important to mention the results of a study published in 2012 in *Human Reproduction*. This study, which was conducted by researchers in Italy, examined the use of topical almond oil and other products during pregnancy. The researchers found that pregnant women who were regular users of topical almond oil had higher rates of preterm births. (Regular use was defined as at least three months of the pregnancy.) "The daily application of almond oil to the abdomen to alleviate stretch marks was significantly associated with the occurrence of PTB [preterm birth], even after controlling for confounding factors such as smoking habit and multiple pregnancies."[9]

NOTES

1. Ahmad, Zeeshan. February 2010. "The Uses and Properties of Almond Oil." *Complementary Therapies in Clinical Practice* 16(1): 10–12.

2. Missouri University of Science and Technology. www.mst.edu.

3. Mori, A. M., R. V. Considine, and R. D. Mattes. January 2011. "Acute and Second-Meal Effects of Almond Form in Impaired Glucose Tolerant Adults: A Randomized Crossover Trial." *Nutrition & Metabolism* 28(8): 6.

4. Jia, Xiao-Yan, Qing-An Zhang, Zhi-Qi Zhang et al. March 15, 2011. "Hepatoprotective Effects of Almond Oil Against Carbon Tetrachloride Induced Liver Injury in Rats." *Food Chemistry* 125(2): 673–678.

5. Sultana, Y., K. Kohli, M. Athar et al. March 2007. "Effect of Pre-Treatment of Almond Oil on Ultraviolet B-Induced Cutaneous Photoaging in Mice." *Journal of Cosmetic Dermatology* 6(1): 14–19.

6. Vaivre-Douret, L., D. Oriot, P. Blossier et al. January 2009. "The Effect of Multimodel Stimulation and Cutaneous Application of Vegetable Oils on Neonatal Development in Preterm Infants: A Randomized Controlled Trial." *Child: Care Health and Development* 35(1): 96–105.

7. Saidi, H. and S. Shojaie. May 2012. "Effect of Sweet Almond Oil on Survival Rate and Plasma Cholinesterase Activity of Aluminum Phosphide-Intoxicated Rats." *Human and Experimental Toxicology* 31(5): 518–522.

8. Taşhan, Sermin Timur and Ayşe Kafkasli. June 2012. "The Effect of Bitter Almond Oil and Massaging on Striae Gravidarum in Primiparaous Women." *Journal of Clinical Nursing* 21(11–12): 1570–1576.

9. Facchinetti, F., G. Pedrielli, Benoni et al. November 2012. "Herbal Supplements in Pregnancy: Unexpected Results from a Multicentre Study." *Human Reproduction* 27(11): 3161–3167.

REFERENCES AND RESOURCES

Magazines, Journals, and Newspapers

Ahmad, Zeeshan. February 2010. "The Uses and Properties of Almond Oil." *Complementary Therapies in Clinical Practice* 16(1): 10–12.

Facchinetti, F., G. Pedrielli, G. Benoni et al. November 2012. "Herbal Supplements in Pregnancy: Unexpected Results from a Multicentre Study." *Human Reproduction* 27(11): 3161–3167.

Jia, Xiao-Yan, Qing-An Zhang, Zhi-Qi Zhang et al. March 15, 2011. "Hepatoprotective Effects of Almond Oil Against Carbon Tetrachloride Induced Liver Injury in Rats." *Food Chemistry* 125(2): 673–678.

Mori, Alisa M., Robert V. Considine, and Richard D. Mattes. January 28, 2011. "Acute and Second-Meal Effects of Almond Form in Impaired Glucose Tolerant Adults: A Randomized Crossover Trial." *Nutrition & Metabolism* 8(1): 6.

Saidi, H. and S. Shojaie. May 2012. "Effect of Sweet Almond Oil on Survival Rate and Plasma Cholinesterase Activity of Aluminum Phosphide-Intoxicated Rats." *Human & Experimental Toxicology* 31(5): 518–522.

Sultana, Yasmin, Kanchan Kohli, M. Athar et al. March 2007. "Effect of Pre-treatment of Almond Oil on Ultraviolet N-Induced Cutaneous Photoaging in Mice." *Journal of Cosmetic Dermatology* 6(1): 14–19.

Taşhan, Sermin Timur and Ayşe Kafkasli. June 2012. "The Effect of Bitter Almond Oil and Massaging on Striae Gravidarum in Primiparaous Women." *Journal of Clinical Nursing* 21(11–12): 1570–1576.

Vaivre-Douret, L., D. Oriot, P. Blossier et al. January 2009. "The Effect of Multimodal Stimulation and Cutaneous Application of Vegetable Oils on Neonatal Development in Preterm Infants: A Randomized Controlled Trial." *Child: Care, Health, and Development* 35(1): 96–105.

Website

Missouri University of Science and Technology. www.mst.edu.

Argan Oil

Until fairly recent times, not many people had heard of argan oil. After all, almost all of the argan trees are grown in the barren land of southwest Morocco. And, traditionally, the preparation of the oil was a laborious process, so supplies of argan oil were somewhat limited. It was used primarily in Morocco as a topical oil that treats dry skin, acne, psoriasis, eczema, wrinkles, joint pain, and skin inflammation.

More recently, technology has improved the production of argan oil. It is more readily available. In addition, argan oil is now known to contain oleic acid, a monounsaturated fatty acid that is thought to have numerous therapeutic properties. It also contains tocopherals, which are molecules with strong antioxidant and free radical scavenging properties. While people continue to apply argan oil to their skin, they also consume it. Argan oil is believed to support cardiovascular health and destroy cancer cells; it may even help prevent diabetes.[1]

Relatively expensive, pure argan oil may be available in some specialty stores, but it is easier to buy online. More readily available is argan oil added to other products, such as shampoo with argan oil.

SUPPORTS CARDIOVASCULAR HEALTH

In a study published in 2012 in the *British Journal of Nutrition*, researchers from Algeria, France, and Spain investigated the potential cardiovascular benefits of argan oil. The cohort consisted of 40 healthy subjects between the ages of 25 and 45; everyone had a normal body mass index and normal blood pressure reading. For four weeks, 20 subjects ate a breakfast consisting of 15g/day of argan oil with toasted bread; the 20 control subjects ate their normal breakfast meals and no argan oil. The researchers found that the daily consumption of argan oil supported a number of markers that improve cardiovascular health. For example, the researchers learned that the consumption of argan oil lowered levels of low density lipoprotein (LDL or "bad") cholesterol and total cholesterol. The researchers concluded that argan oil "is able to positively modulate some surrogate markers of CVD [cardiovascular disease]."[2]

In a study published in 2012 in *Nutrition*, researchers based in Morocco investigated the antithrombotic activity of argan oil (the ability of argan oil to prevent blood clots). The researchers began by randomly dividing mice into three groups. The mice in one group were treated with distilled water; they served as controls. The mice in the second group were treated with argan oil; and the mice in the third group were treated with acetyl salicylic acid (aspirin), an antithrombotic medication. The trial continued for one week. One hour after the final treatment, the researchers induced pulmonary thrombosis in the mice. The researchers found that argan oil offered a degree of protection against the formation of blood clots. Moreover, "this antithrombotic effect was well highlighted by its protection against the paralysis or death of mice and by the histologic lung examination, where totally and partly occluded blood vessel numbers were decreased in the oil-treated group." The mice in the aspirin treatment group did almost as well as the argan oil–treated mice; the mice in the distilled water group did poorly. The researchers concluded that argan oil may now "be considered a new natural dietary source for the nutritional prevention of cardiovascular disorders."[3]

It is well known that people with type 2 diabetes are at increased risk for cardiovascular illness. Would the consumption of argan oil lower that risk? In a study published in 2011 in the *International Journal of Endocrinology*, researchers from Morocco assembled 86 subjects with type 2 diabetes between the ages of 40 and 80. All of the subjects had abnormal serum lipid levels. For the first two weeks of the trial, the subjects consumed 20g/day of butter with toasted bread for breakfast. During the next three weeks of the study, the subjects were randomized into two groups. One group of 43 subjects continued to eat butter with breakfast. A second group, which also had 43 subjects, consumed 25mL/day of argan oil with breakfast. By the end of the trial, 41 subjects had left the study. Still, the findings were notable. After three weeks on argan oil, the subjects experienced significant improvements in their lipids. In so doing, they helped prevent the formation of plaques in arteries, thereby improving cardiovascular health.[4]

KILLS PROSTATE CANCER CELLS

In a study published in 2007 in *Cancer Detection and Prevention*, researchers from Morocco and France examined the ability of polyphenols and sterols obtained from argan oil to kill human prostate cells in three different cell lines. The researchers found that the polyphenols and sterols killed prostate cancer cells in all three lines.

The researchers commented that their "results showed that sterols of argan oil produced a time and concentrated-dependent decrease in the number of viable cells in all three tested cell lines." The "data suggest that argan oil may be interesting in the development of new strategies for prostate cancer prevention." Still the researchers cautioned that more studies are needed. "Before drawing final conclusions about the mechanisms of the anticancer effects of polyphenols and sterols extracted from argan oil, further investigations concerning the cell proliferation and apoptsis would be necessary."[5]

MAY HELP PREVENT TYPE 1 DIABETES

In a two-part study published in 2012 in *Phytotherapy Research*, researchers from Morocco wanted to learn more about the ability of argan oil to prevent the development of diabetes in Wistar rats. In the first part of the trial, healthy rats were divided into three groups of six rats. The rats were fed either distilled water, virgin argan oil, or glibenclamide, a diabetes medication. Their blood glucose levels were assessed before the experiment and at various times after the experiment.

During the second part of the trial, which lasted for several days, rats were chemically induced to develop type 1 diabetes. Some of these rats were pretreated with virgin argan oil; others were induced to develop diabetes while simultaneously receiving virgin argan oil. The rats in the control group were induced to develop diabetes, but they had only water. Another group of rats, also induced with diabetes, was treated with table oil.

The researchers found that the consumption of argan oil by healthy rats did not result in "any significant change in blood sugar levels." On the other hand, when they consumed glibenclamide, the medication "caused a significant reduction of blood glucose levels." According to the researchers, "one of the cytoprotective mechanisms of argan oil against chemically induced diabetes in rats may be the enhancement of the antioxidant status of the tissues."[6]

MAY BE VALUABLE FOR THOSE WITH
OBESITY-LINKED INSULIN RESISTANCE

In a study published in 2009 in *Metabolism Clinical and Experimental*, researchers from Canada, France, and Morocco wanted to learn if argan oil or fish oil would have any metabolic benefits in a dietary model of obesity-linked insulin resistance. They began by dividing their male Wistar rats into four groups. The rats in one group were fed regular rat food; they served as the controls. The rats in a second group were fed a high-fat/high-sucrose diet. The rats in the third group were fed a high-fat/high-sucrose diet and 6% of the fat was replaced with argan oil. The rats in the fourth group had a high-fat/high-sucrose diet and 6% of the fat was replaced with fish oil. The researchers found that the intake of either argan oil or fish oil "can improve certain features of the metabolic syndrome in the HFHS-fed [high-fat/high-sucrose fed] rat, a well-known model of diet-induced insulin resistance." For example, the rats fed argan oil "demonstrated a marked improvement in fasting plasma glucose [blood sugar levels]." And, rats fed argan oil had "significantly enhanced insulin sensitivity in both [the] fat and liver." The researchers concluded that diets containing argan oil or fish oil may help prevent some of the "deleterious effects" of eating a high-fat/high-sugar diet. These two oils may prove to be "promising nutritional tools to combat insulin resistance and associated comorbidities."[7]

IS ARGAN OIL BENEFICIAL?

Though there is obviously a need for more research on argan oil, it appears to have a number of useful properties. In general, it should not be heated; it may be drizzled over salads and other prepared foods. While argan oil is usually well tolerated, there is a report in the literature of a 34-year-old Moroccan man who experienced an anaphylaxis response. Fortunately, this appears to be an exceedingly rare event.[8]

NOTES

1. Guillaume, Dom and Zoubida Charrouf. September 2011. "Argan Oil." *Alternative Medicine Review* 16(3): 275–279.

2. Sour, Souad, Meriem Belarbi, Darine Khaldi et al. June 2012. "Argan Oil Improves Surrogate Markers of CVD in Humans." *British Journal of Nutrition* 107(12): 1800–1805.

3. Mekhfi, Hassane, Fatima Belmekki, Abderrahim Ziyyat et al. September 2012. "Antithrombotic Activity of Argan Oil: An *in vivo* Experimental Study." *Nutrition* 28(9): 937–941.

4. Ould Mohamedou, M. M., K. Zouirech, M. El Messal et al. 2011. "Argan Oil Exerts an Antiatherogenic Effect by Improving Lipids and Susceptibility of LDL to Oxidation in Type 2 Diabetes Patients." *International Journal of Endocrinology* Article ID 747835: 8 pages.

5. Bennani, H., A. Drissi, F. Giton et al. 2007. "Antiproliferative Effect of Polyphenols and Sterols of Virgin Argan Oil on Human Prostate Cancer Cell Lines." *Cancer Detection and Prevention* 31(1): 64–69.

6. Ibid.

7. Samane, Samira, Raymond Christon, Luce Dombrowski et al. July 2009. "Fish Oil and Argan Oil Intake Differently Modulate Insulin Resistance and Glucose Intolerance in a Rat Model of Dietary-Induced Obesity." *Metabolism Clinical and Experimental* 58(7): 909–919.

8. Astier, C., Y. El Alaoui Benchad, D.-A. Moneret-Vautrin et al. May 2010. "Anaphylaxis to Argan Oil." *Allergy* 65(5): 662, 663.

REFERENCES AND RESOURCES

Magazines, Journals, and Newspapers

Astier, C., Y. El Alaoui Benchad, D.-A. Moneret-Vautrin et al. May 2010. "Anaphylaxis to Argan Oil." *Allergy* 65(5): 662, 663.

Bellahcen, Said, Hassane Mekhfi, Abderrahim Ziyyat et al. February 2012. "Prevention of Chemically Induced Diabetes Mellitus in Experimental Animals by Virgin Argan Oil." *Phytotherapy Research* 26(2): 180–185.

Bennani, H., A. Drissi, F. Giton et al. 2007. "Antiproliferative Effect of Polyphenols and Sterols of Virgin Argan Oil on Human Prostate Cancer Cell Lines." *Cancer Detection and Prevention* 31(1): 64–69.

Guillaume, Dom and Zoubida Charrouf. September 2011. "Argan Oil." *Alternative Medicine Review* 16(3): 275–279.

Mekhfi, Hassane, Fatima Belmekki, Abderrahim Ziyyat et al. September 2012. "Antithrombotic Activity of Argan Oil: An *in vivo* Experimental Study." *Nutrition* 28(9): 937–941.

Ould Mohamedou, M.M., K. Zouirech, M. El Messal et al. 2011. "Argan Oil Exerts an Antiatherogenic Effect by Improving Lipids and Susceptibility of LDL to Oxidation in Type 2 Diabetes Patients." *International Journal of Endocrinology* Article ID 747835: 8 pages.

Samane, Samira, Raymond Christon, Luce Dombrowski et al. July 2009. "Fish Oil and Argan Oil Intake Differently Modulate insulin Resistance and Glucose Intolerance in a Rat Model of Dietary-Induced Obesity." *Metabolism Dietary and Experimental* 58(7): 909–919.

Sour, Souad, Meriem Belarbi, Darine Khaldi et al. June 2012. "Argan Oil Improves Surrogate Markers of CVD in Humans." *British Journal of Nutrition* 107(12): 1800–1805.

Website

Argan Oils. http://arganoils.com.

Avocado Oil

Because they contained such high amounts of fat, for many years, large numbers of people avoided consuming avocadoes and avocado oil. That was unfortunate. Avocadoes and avocado oil have a number of healthful properties. They have high amounts of vitamin E and unsaturated fats; they have more protein than any fruit and more potassium than bananas. Avocados may also play a role in supporting cardiovascular health and preventing cancer.

It is not always easy to find avocado oil in traditional markets. But, it is available in specialty stores and online. However, avocado oil tends to be pricy.

MAY SUPPORT CARDIOVASCULAR HEALTH

In a study published in 2005 in the *Journal of Ethnopharmacology*, researchers from Mexico compared the blood pressure readings of male Wistar rats fed a diet rich in avocado oil to male Wistar rats fed a control diet. The animals were randomly divided into two groups of five rats each. One group was fed a 10% avocado oil and lab chow diet for two weeks; the other group was fed lab chow. The researchers found that a diet rich in avocado oil altered levels of essential fatty acids in the kidneys. This, in turn, changed the way the kidneys responded to hormones that regulate blood pressure. Moreover, the avocado oil rich diet also influenced the fatty acids in the heart, which further improved blood pressure. The researchers concluded that an "avocado oil-rich diet modifies the fatty acid content in cardiac and renal membranes in a tissue-specific manner."[1]

In a study published in 2007 in Spanish in *Archivos de Cardiología de México*, researchers from Mexico wanted to learn the effects of avocado consumption on the cardiovascular health of male Wistar rats. The cohort of 30 rats was divided into two groups. Avocado was added to the diets of 15 rats; the other 15 rats ate their regular rat food. After five weeks, when compared to the rats in the control group, the rats that ate avocado experienced about a 27% reduction in their triglyceride levels and increases of 17% in their high-density lipoproteins (HDL or "good" cholesterol).[2]

MAY IMPROVE ABSORPTION OF NUTRIENTS

In a study published in 2005 in *The Journal of Nutrition*, researchers from Columbus, Ohio (United States), noted that it is believed that carotenoids are more readily absorbed when they are consumed in combination with dietary lipids (a broad group of naturally occurring molecules that include fat, waxes, sterols, and fat-soluble vitamins). However, since carotenoid-rich foods tend to contain very few lipids, the researchers wanted to learn if adding avocado oil (as a source of lipids) would improve the absorption of carotenoids in humans. Their cohort consisted of 11 healthy subjects between the ages of 21 and 42. During the study, the subjects ate salads with and without avocados and avocado oil. The researchers found that "the carotenoid absorption from salad with the addition of either avocado or avocado oil was greater than from salad alone." Moreover, "the high dose of avocado (150g) and the fat-equivalent avocado oil (24g) both increased the absorption from salad of all carotenoids studied."[3]

MAY PROTECT THE SKIN AGAINST
UV-INDUCED DAMAGE

In a study published in 2011 in the *Archives of Dermatological Research*, researchers from Israel, Cambridge, Massachusetts (United States), and New York City investigated whether polyhydroxylated fatty alcohols, which are derived from avocado oil, could protect skin from ultraviolet (UV)-induced damage. Such damage has been associated with aging and skin cancer. During the trial, some samples of human skin were treated with polyhydroxylated fatty alcohols.

Other samples remained untreated. The samples were then exposed to a type of UV radiation known as UVB. (Composed of shorter wavelengths, UVB primarily affects the top layer of skin known as the epidermis.) The researchers found that avocado oil protected the skin from potential problems caused by UVB. They concluded that "the results of the present study provide evidence that PFA protect skin cells from UVB-induced damage, enhance the DNA damage repair mechanism, and inhibit inflammation. Consequently, PFA treatment may be beneficial in the photo-protection of the skin."[4]

MAY BE USEFUL FOR WOUND HEALING

In a study published in 2008 in the *Journal of Wound Healing*, researchers based in Augustine, Trinidad, noted that avocado oil is rich in "nutrient waxes, proteins and minerals, as well as vitamins A, D and E." Moreover, they wrote, it is "an excellent source of enrichment for dry, damaged or chapped skin." Would avocado also demonstrate wound-healing properties? To learn if it could, the researchers divided rats into four groups of five each. After receiving cuts to their skin, the rats were treated with either topical or oral 300mg/kg/day of avocado extract. The rats in the control group received no avocado treatment. The degree of wound healing was determined by the levels of wound closing and the amount of wound repair molecules, known as hydroxyproline, in the wound tissue. The researchers found that the wounds of the rats that received oral or topical treatments healed in 14 days. And, the wounds of the rats in the control group required 17 days to heal. They concluded that avocado may play a role in wound healing.[5]

MAY HELP THE MILLIONS DEALING
WITH OSTEOARTHRITIS

In a prospective, double-blind, randomized, study published in 2010 in *Clinical Rheumatology*, researchers from the Czech Republic, France, and Hungary compared two different treatments for osteoarthritis, a degenerative joint disease characterized by damage to cartilage collagen. The initial cohort consisted of 364 people who had dealt with osteoarthritis of the knees for longer than six months and experienced pain and functional discomfort. Of these, it was determined that 361 were eligible for evaluation. All the subjects were at least 45 years old. For six months, one group of 183 people took the avocado and soybean unsaponifiable supplement (Piascledine, 300mg) each day; the second group, which consisted of 178 people, took chondroitin sulfate three times a day. (Chondroitin sulfate is known to ease the symptoms of osteoarthritis.) After the trial was complete, the subjects were followed for an additional two months. The researchers found that both groups received relief from their osteoarthritis symptoms. "Pain on active movement, pain at rest and global assessment of knee osteoarthritis decreased in both groups to approximately 50% of the initial value at the end of therapy, and remained unchanged until the end of the post-treatment period." The participants in both groups were able to reduce their intake of medications for the symptoms of osteoarthritis.[6]

MAY BE USEFUL FOR THE SYMPTOMS OF PSORIASIS

Psoriasis is a common skin condition that causes skin redness and irritation. While there are many over-the-counter and prescriptive medications that are used to treat psoriasis, large numbers of people have trouble finding one that consistently works well. And, some treatments have troubling side effects, making it less likely that patients will comply. In a randomized, prospective trial published in 2001 in *Dermatology,* researchers from Germany wondered if a cream containing vitamin B12 and avocado oil would be useful for psoriasis. The cohort consisted of 10 men and 3 women between the ages of 38 and 67; everyone had chronic plaque psoriasis. For 12 weeks, twice daily, the subjects applied the B12 and avocado cream or a cream containing calcipotriol, a synthetic derivative of vitamin D that is known to be useful for psoriasis, to their psoriatic plaques. The plaques were evaluated at weeks 2, 4, 8, and 12. Because of a failure to adhere to the protocol, two people were excluded from the final analysis. Still, at the end of the trial, the researchers found "no significant differences in [the] treatment effect of calcipotriol and the vitamin B12 cream containing avocado oil." Interestingly, after four weeks of treatment, "there was a marked diminution in the efficacy of calcipotriol while the frequency and severity of skin irritation increased, whereas the efficacy of the vitamin B12 cream containing avocado oil remained largely constant over the whole observation period."[7]

IS AVOCADO OIL BENEFICIAL?

Clearly, avocado oil appears to have a number of benefits. For most people, it may well be a useful addition to the diet. For example, since avocado oil is tolerant of high heat, it can easily be used as an all-purpose cooking oil. However, a 2009 article in the *International Journal of Food Sciences and Nutrition* mentioned that the shelf life of avocado oil may be less than that indicated on the label.[8] Once opened, it is a good idea to store avocado oil in the refrigerator.

NOTES

1. Salazar, M. J., M. El Hafidi, G. Pastelin et al. 2005. "Effect of an Avocado Oil-Rich Diet over an Angiotensin II-Induced Blood Pressure Response." *Journal of Ethnopharmacology* 98(3): 335–338.

2. Pérez Méndez, Óscar and Lizbeth García Hernández. January–March 2007. "High-Density Lipoproteins (HDL) Size and Composition Are Modified in the Rat by a Diet Supplemented with 'Hass' Avocado (*Persea americana* Miller)." *Archivos de Cardiología de México* 77(1): 17–24.

3. Unlu, Nuray Z., Torsten Bohn, Steven K. Clinton, and Steven J. Schwartz. March 1, 2005. "Carotenoid Absorption from Salad and Salsa by Humans Is Enhanced by the Addition of Avocado or Avocado Oil." *The Journal of Nutrition* 135(3): 431–436.

4. Rosenblat, Gennady, Shai Meretski, Joseph Segal et al. 2011. "Polyhydroxylated Fatty Alcohols Derived from Avocado Suppress Inflammatory Response and Provide Non-Sunscreen Protection against UV-Induced Damage in Skin Cells." *Archives of Dermatological Research* 303(4): 239–246.

5. Nayak, B. S., S. S. Raju, and A. V. Chalapathi Rao. March 2008. "Wound Healing Activity of *Persea americana* (Avocado) Fruit: A Preclinical Study on Rats." *Journal of Wound Care* 17(3): 123–126.

6. Pavelka Karel, Philippe Cost, Pál Géher, and Gerhard Krejci. 2010. "Efficacy and Safety of Piascledine 300 versus Chondroitin Sulfate in a Six Months Treatment plus Two Months Observation in Patients with Osteoarthritis of the Knee." *Clinical Rheumatology* 29(6): 659–670.

7. Stücker, Markus, Ulrike Memmel, Matthias Hoffmann et al. 2001. "Vitamin B12 Cream Containing Avocado Oil in the Therapy of Plaque Psoriasis." *Dermatology* 203(2): 141–147.

8. Kochhar, S. Parkash and C. Jeya Henry. September 2009. "Oxidative Stability and Shelf-Life Evaluation of Selected Culinary Oils." *International Journal of Food Sciences and Nutrition* 60(S7): 289–296.

REFERENCES AND RESOURCES

Magazines, Journals, and Newspapers

Kochhar, S. Parkash and C. Jeya Henry. September 2009. "Oxidative Stability and Shelf-Life Evaluation of Selected Culinary Oils." *International Journal of Food Sciences and Nutrition* 60(S7): 289–296.

Nayak, B. S., S. S. Raju, and A. V. Chalapathi Rao. March 2008. "Wound Healing Activity of *Persea americana* (Avocado) Fruit: A Preclinical Study on Rats." *Journal of Wound Care* 17(3): 123–126.

Pavelka, Karel, Philippe Coste, Pál Géher, and Gerhard Krejci. 2010. "Efficacy and Safety of Piascledine 300 versus Chondroitin Sulfate in a Six Months Treatment plus Two Months of Observation in Patients with Osteoarthritis of the Knee." *Clinical Rheumatology* 29(6): 659–670.

Pérez Méndez, Óscar and Lizbeth García Hernández. January–March 2007. "High-Density Lipoproteins (HDL) Size and Composition Are Modified in the Rat by a Diet Supplemented with 'Hass' Avocado (*Persea americana* Miller)." *Archivos de Cardiología de México* 77(1): 17–24.

Rosenblat, Gennady, Shai Meretski, Joseph Segal et al. 2011. "Polyhydroxylated Fatty Alcohols Derived from Avocado Suppress Inflammatory Response and Provide Non-Sunscreen Protection against UV-Induced Damage in Skin Cells." *Archives of Dermatological Research* 303(4): 239–246.

Salazar, M. J., M. El Hafidi, G. Pastelin et al. 2005. "Effect of an Avocado Oil-Rich Diet over an Angiotensin II-Induced Blood Pressure Response." *Journal of Ethnopharmacology* 98(3): 335–338.

Stücker, Markus, Ulrike Memmel, Matthias Hoffmann et al. 2001. "Vitamin B12 Cream Containing Avocado Oil in the Therapy of Plaque Psoriasis." *Dermatology* 203(2): 141–147.

Unlu, Nuray Z., Torsten Bohn, Steven K. Clinton, and Steven J. Schwartz. March 1, 2005. "Carotenoid Absorption from Salad and Salsa by Humans Is Enhanced by the Addition of Avocado or Avocado Oil." *The Journal of Nutrition* 135(3): 431–436.

Website

LIVESTRONG.COM. http://livestrong.com.

Borage Seed Oil

Native to Syria, borage is now grown throughout the Mediterranean, Middle East, North Africa, Europe, and South America. The oil is made from the seeds of the plant. It is a potent source of gamma-linolenic acid (GLA), a good omega-6 fat.

While research on borage seed oil is still somewhat limited, borage seed oil is thought to have anti-inflammatory properties. It is also said to be useful for a number of skin problems such as eczema, psoriasis, and rashes.

Borage seed oil is sold as a supplement. Though readily available online, it may be harder to locate in conventional stores. However, it should be sold at well-stocked health food stores. Good quality borage seed oil is moderately priced. Additionally, borage seed oil may be an ingredient in skin care products. Readily available in drug stores, giant chain stores, and online, these products tend to be moderately priced.

MAY BE USEFUL FOR PEOPLE WITH RHEUMATOID ARTHRITIS

In a randomized, double-blind study published in 2011 in *Evidence-Based Complementary and Alternative Medicine*, researchers from the University of Massachusetts Medical School in Worcester, Massachusetts, commented that people with rheumatoid arthritis (RA) have an increased risk of cardiovascular disease. (In RA, the immune system attacks the lining of the joints. It makes the joints swollen, stiff, and painful.) So, they wanted to determine if treating people with RA with a combination of fish oil and borage seed oil was more effective in controlling elevated lipids levels than treating them with either borage seed oil or fish oil. The initial cohort consisted of 156 people who had a mean age of 59 and who had RA with "active joint inflammation." For 18 months, they took supplements containing borage seed oil, fish oil, or both borage seed oil and fish oil. The researchers found "no significant differences . . . between the three groups: All three treatment groups exhibited similar meaningful improvement in the lipid profile at 9 and 18 months."[1]

In an analysis published in 2011 in the *Cochrane Database of Systematic Reviews*, researchers from Australia and Ann Arbor, Michigan examined 22 randomized and controlled trials on the use of herbal therapy for treating RA. They found that three oils—borage seed, evening primrose, and blackcurrent seed—appeared to reduce the pain associated with RA. The researchers commented that "there is moderate evidence that oils containing GLA (evening primrose, borage, or blackcurrent seed oil) afford some benefit in relieving symptoms for RA."[2]

PREVENTING OSTEOPOROSIS

In a study published in 2012 in *Bone*, researchers from France examined the supplementation of various oils in mice bred to develop a form of senile osteoporosis. "This model exhibits both sarcopenia [degenerative loss of skeletal muscle mass and strength associated with aging] and osteoporosis and matches clinical features observed in elderly." The researchers divided the mice into several groups and fed them the following diets: standard diet, sunflower diet, borage seed diet, and fish oil. After 12 months, the mice were sacrificed and their tissues were examined. Given the very high rates of osteoporosis among the elderly, the results are important. The researchers found that "both borage and fish oil enriched diets were able to counteract aging-associated bone loss, mostly by preventing inflammation establishment and its consequences on bone cell behavior." They concluded that "this encouraging data suggest that nutritional approaches are relevant regarding osteoporosis and offer promising alternative in the design of new strategies for preventing aged-related locomotor dysfunctions."[3]

HELPS PREVENT FORMERLY OBESE PEOPLE
FROM REGAINING WEIGHT

In a double-blind study published in 2007 in *The Journal of Nutrition*, researchers from the University of California at Davis attempted to learn if the supplementation of the diet with borage would suppress weight regain in obese men and women who had lost large amounts of weight. Fifty formerly obese people were randomly placed in one of two groups. The members of one group received borage supplementation; the members of the control group took olive oil. After 12 people in each group had completed one year of supplementation, the researchers found that those in the control group had regained much more weight than those on borage oil. At that point, "in recognition that weight regain is not benign," the researchers terminated the study. The researchers concluded that borage seed oil "reduced weight regain in humans following major weight loss, suggesting a role for essential fatty acids in fuel partitioning in humans prone to obesity."[4]

IMPROVES SKIN CONDITIONS

In a 12-week, double-blind, placebo-controlled study published in 2009 in the *British Journal of Nutrition*, researchers from Germany and France wanted to determine if borage seed oil and flaxseed oil could improve skin conditions in women. They recruited 45 female healthy nonsmoking women between the ages of 18 and 65; all the women had sensitive, dry skin. The women were then divided into three groups of 15. One group of women took borage seed oil; a second group of women took flaxseed oil; and a third group of women took placebos. In order to achieve skin irritation, nicotinate was applied to the skin. The researchers found that both borage seed and flaxseed oils diminished skin reddening, increased skin hydration, and reduced skin roughness and irritation. The researchers noted that their study demonstrated that "dietary intervention is a way of improving skin conditions and providing protection." However, the "effects are moderate and develop over a long-term period."[5]

In a report published in 2007 in the *European Journal of Dermatology*, researchers based in Poland noted that borage seed oil supplementation has been used effectively for people suffering from atopic dermatitis, a long-term skin disorder in which there are scaly and itchy rashes. So, they decided to investigate whether wearing undershirts coated with borage seed oil for four weeks would help children with atopic dermatitis. Their cohort consisted of 26 children between the ages of 2 and 7, with mild to moderate atopic dermatitis. "The borage oil was gradually released from the cotton fibers and absorbed into the skin." By the end of the study, the majority of children experienced improvements in their symptoms.

ANTIOXIDANT ACTIVITY

In a study published in 2011 in *Food and Nutrition Sciences*, researchers from India evaluated the antioxidant activity of borage, oregano, and ajowan extracts. Why is this important? Antioxidants "are reported to play a critical role in delay of the onset as well as prevention of many chronic degenerative diseases." The researchers found that all three extracts were "potential dietary sources of natural antioxidants" and they have "the potential to be used as natural food antioxidants in place of synthetic additives."[6]

IS BORAGE SEED OIL BENEFICIAL?

Borage seed oil appears to have a number of positive aspects. And, many people may wish to include it in their supplementation. Still, information from an article that appeared in 2011 in the *Journal of Medical Toxicology* should be mentioned. Medical professionals from Atlanta, Georgia, and Kansas City, Missouri, described a 41-year-old "previously healthy female" who arrived at a medical facility "with continuous seizure activity." After carefully analyzing her history, medications, and supplements, the medical professionals concluded that the only logical cause of her seizures was borage seed oil. They noted that "herbal products and alternative medicines should be considered in the differential diagnosis of status epilepticus."[7]

NOTES

1. Olendzki, Barbara C., Katherine Leung, Susan Van Buskirk et al. 2011. "Treatment of Rheumatoid Arthritis with Marine and Botanical Oils: Influence on Serum Lipids." *Evidence-Based Complementary and Alternative Medicine*. Article ID 827286: 9 pages.

2. Cameron, M., J. J. Gagnier, and S. Chrubasik. February 16, 2011. "Herbal Therapy for Treating Rheumatoid Arthritis." *Cochrane Database of Systematic Reviews* 2. Article Number CD002948.

3. Wauquier, Fabien, Valentin Barquissau, Laurent Léotoing et al. 2012. "Borage and Fish Oils Lifelong Supplementation Decreases Inflammation and Improves Bone Health in a Murine Model of Senile Osteoporosis." *Bone* 50(2): 553–561.

4. Schirmer, Marie A. and Stephen D. Phinney. June 2007. "Y-Linolenate Reduces Weight Regain in Formerly Obese Humans." *The Journal of Nutrition* 137(6): 1430–1435.

5. De Spirt, Silke, Wilhelm Stahl, Hagen Tronnier et al. 2009. "Intervention with Flaxseed and Borage Oil Supplements Modulates Skin Condition in Women." *British Journal of Nutrition* 101(3): 440–445.

6. Khanum, Hafeeza, Kulathooran Ramalaksmi, Pullabhatla Srinivas et al. July 2011. "Synergistic Antioxidant Action of Oregano, Ajowan and Borage Extracts." *Food and Nutrition Sciences* 2(5): 387–392.

7. Al-Khamees, W. A., M. D. Schwartz, S. Alrashdi et al. 2011. "Status Epilepticus Associated with Borage Oil Ingestion." *Journal of Medical Toxicology* 7(2):154–157.

REFERENCES AND RESOURCES
Journals, Magazines, and Newspapers

Al-Khamees, W. A., M. D. Schwartz, S. Alrashdi et al. June 2011. "Status Epilepticus Associated with Borage Oil Ingestion." *Journal of Medical Toxicology* 7(2): 154–157.

Baumann, Leslie S. September 2011. "Borage Seed Oil." *Skin & Allergy News* 42(9): 22.

Cameron, M., J. J. Gagnier, and S. Chrubasik. February 16, 2011. "Herbal Therapy for Treating Rheumatoid Arthritis." *Cochrane Database of Systematic Reviews* 2. Article Number CD002948.

De Spirt, Silke, Wilhelm Stahl, Hagen Tronnier et al. 2009. "Intervention with Flaxseed and Borage Oil Supplements Modulates Skin Condition in Women." *British Journal of Nutrition* 101(3): 440–445.

Kanehara, S., T. Ohtani, K. Uede, and F. Furukawa. September–October 2007. "Undershirts Coated with Borage Oil Alleviate the Symptoms of Atopic Dermatitis in Children." *European Journal of Dermatology* 17(5): 448, 449.

Khanum, Hafeeze, Kulathooran Ramalakshmi, Pullabhatla Srinivas el al. July 2011. "Synergistic Antioxidant Action of Oregano, Ajowan, and Borage Extracts." *Food and Nutrition Sciences* 2(5): 387–392.

Olendzki, Barbara C., Katherine Leung, Susan Van Buskirk et al. 2011. "Treatment of Rheumatoid Arthritis with Marine and Botanical Oils: Influence on Serum Lipids." *Evidence-Based Complementary and Alternative Medicine*. Article ID 827286: 9 pages.

Schirmer, Marie A. and Stephen D. Phinney. June 2007. "Y-Linolenate Reduces Weight Regain in Formerly Obese Humans." *The Journal of Nutrition* 137(6): 1430–1435.

Wauquier, Fabien, Valentin Barquissau, Laurent Léotoing et al. 2012. "Borage and Fish Oils Lifelong Supplementation Decreases Inflammation and Improves Bone Health in a Murine Model of Senile Osteoporosis." *Bone* 50(2): 553–561.

Website

Skin & Allergy News. www.skinandallergynews.com.

Canola Oil

Canola oil is derived from the crushed seeds of the canola plant, which is a member of the Brassica family that also includes cauliflower, cabbages, and broccoli.

When mature, canola plants are between three and six feet tall; they have yellow flowers and pods that resemble pea pods. Each pod contains about 20 tiny seeds that are either brownish yellow or black. According to the U.S. Canola Association, "canola oil is one of the healthiest, most versatile cooking oils in the world." It is high in monounsaturated fat and omega-3 fat and low in saturated fat; it is a good source of vitamins E and K. In addition, "canola oil is second highest in plant sterols of all vegetable oils." Plant sterols lower the risk of cardiovascular disease.[1]

Canola oil is widely available in supermarkets, specialty stores, and online. Generally, it is moderately priced. Canola oil is used in a wide variety of cooking throughout the world. It is the number one vegetable oil in Canada and Japan and the number two vegetable oil in the United States and Mexico.[2]

PROVIDES PROTECTION FROM CANCER

In a study published in 2011 in *Nutrition and Cancer*, researchers from South Dakota, Pennsylvania, and China examined the ability of canola oil to reduce the size and incidence of colon cancer. The researchers began with three groups of 24 male Fischer rats. The rats in the first group were fed a regular rat diet; the rats in the second group were fed a rat diet plus 15% corn oil; the rats in the third group were fed a rat diet plus 15% canola oil. After one week on these diets, the researchers induced colon cancer with azoxymethane. At the end of the 30-week trial, the researchers found that "dietary canola oil inhibited 58% of the tumor multiplicity and 90% of the tumor size as compared to the corn oil group." They concluded that "dietary canola oil may be chemopreventive for colon tumor development in Fisher rats."[3]

In a population-based, multiethnic study published in 2008 in *Nutrition and Cancer*, researchers from Boston, Los Angeles, and Stanford, California, assessed the effects of different types of dietary fats on the risk for breast cancer. The cohort consisted of 1,703 women treated for breast cancer and 2,045 controls who participated in a population-based case-control study conducted in the San Francisco Bay area. During the research, the researchers determined the intake of total fat as well as the intake of the different types of fat. They learned that a high fat intake was associated with an increased risk for breast cancer. And, the women who cooked with hydrogenated fats or vegetable/corn oils had a higher rate of breast cancer than those who cooked with canola and olive oils. "Women using olive/canola oil (rich in monounsaturated fat) were at lower risk, similar to those who used no fat for cooking."[4]

In a fascinating study published in 2010 in BMC Cancer, researchers from West Virginia (United States) wanted to learn if a maternal diet that contains canola oil instead of corn oil (thus increasing the amount of omega-3 polyunsaturated fatty acids) reduces the risk of mammary gland cancer in offspring. They began with 20 SV 129 mice; these are mice that are mice bred to develop mammary gland cancer. Ten of the mice consumed a diet that contained 10% canola oil; and 10 consumed a diet that contained 10% corn oil. After two weeks on the diet, the mice were bred. Throughout gestation and nursing, the mice continued to consume their assigned diets. And, the findings were certainly notable. When the offspring were 130 days old, the researchers found that the offspring of the mice that consumed canola oil had significantly fewer tumors per mouse; the tumor incidence was half that of the offspring of the mice that consumed corn oil. Forty days later, when the mice were 170 days old, the offspring of the canola- and corn-fed mice had about the same numbers of tumors.

However, the researchers noted, because the tumors in the offspring of the canola-fed mice grew slower than the corn-fed mice, "the appearance of tumors was delayed, the tumor burden (tumor weight) in the canola exposed mice at 170 days of age was significantly less" than that of the offspring of the corn-fed mice. The researchers commented that the "consumption of canola oil by mothers of the experimental mice delayed mammary gland tumor development in this model." Then, the researchers extrapolated their findings to humans; two important factors emerged. First, they commented that it is very easy for people to add canola oil to their diets. And, second, "the maternal diet can have a life-long influence on the development of breast cancer in the daughter." And, they concluded that "substituting canola oil for the corn oil in the maternal diet may decrease risk for breast cancer in the daughter in addition to providing benefit for the mother."[5]

MAY PROVIDE CARDIOVASCULAR SUPPORT

In a cross-sectional study published in 2008 in The American Journal of Clinical Nutrition, researchers from Iran assessed the difference in a number of cardiovascular-related factors between women who consume partially hydrogenated vegetable oils (PHVOs) and those who consume nonhydrogenated vegetable oils, such as canola oil. The cohort consisted of 486 female teachers between the ages of 40 and 60. According to the researchers, PHVOs are the primary source of dietary fat among Iranians. Yet, they found that "higher in-takes of PHVOs were shown to increase the risk of atherosclerosis, cardiovascu-lar disease, metabolic syndrome, insulin resistance, and diabetes." At the same time, they observed that higher consumption of nonhydrogenated oils, includ-ing canola oil, were associated with markers of better cardiovascular health, such as lower circulating concentrations of C-reactive protein.[6]

In a randomized, single-blind crossover study published in 2011 in the *British Journal of Nutrition*, researchers from Canada evaluated the use of canola oil alone or in combination with flaxseed oil for people with elevated levels of cholesterol. The initial cohort consisted of 14 men and 25 women, between the ages of 18 and 65, who had elevated levels of cholesterol. The study de-sign had three phases; each phase was 28 days. These were separated by four to eight week washout periods when the subjects ate their usual diets. The first diet contained supplemental canola oil; the second diet contained supplemental canola oil and flaxseed oil; the third diet contained a blend of Western diet supplemental oils such as nonsalted butter, extra-virgin olive oil, vegetable lard, and sunflower oil. Thirteen men and 23 women completed the study. The re-searchers found that both canola oil and canola oil combined with flaxseed oil had lipid lowering properties. When compared with the Western oil diet results, both canola oil and the canola oil and flaxseed oil combination lowered total cholesterol and low-density lipoprotein (LDL or "bad") cholesterol. In addi-tion, the canola and flaxseed oils "may further target inflammation and athero-genic pathways."[7]

On the other hand, in a study published in 2011 in *Lipids in Health and Dis-ease*, researchers from Australia placed stroke-prone spontaneously hypertensive (SHRSP) rats on diets containing 10% canola oil or 10% soybean oil. Other than the canola and soybean oils, the rats' diets contained no additional fat. Although the consumption of canola oil lowered levels of total cholesterol and LDL, the researchers found that the consumption of canola oil by the SHRSP rats reduced their life span by about 13%. The researchers commented that "the plasma lipids were reduced after canola oil ingestion highlighting the health benefits of canola oil intake. Despite the improvement in the plasma lipids, canola oil was detri-mental to the SHRSP rat as their lifespan is reduced."[8]

MAY BE USEFUL IN PREVENTING WEIGHT GAIN

In a study published in 2010 in *Nutrition Research*, researchers from Canada wanted to determine if the consumption of canola oil may help prevent weight gain. In two different studies, the researchers fed Sprague-Dawley rats diets that contained canola oil, lard, and butter—low, moderate, and higher sources of saturated fatty acids. The researchers found that the canola-fed animals "ad-justed their intake based on the energy density of the diet." The animals fed lard and butter failed to do this; they exhibited less control over what they ate. In fact, the researchers learned that "obesity is promoted by a diet rich in saturated fats." They concluded that their "results confirmed the hypothesis that a SFA [saturated fatty acids] diet is more obesogenic [supportive of obesity] than diets with lower SFA content."[9] So, people hoping to maintain a health weight should consider using canola oil and avoid lard and butter whenever possible.

SUPPORTS INFANT GROWTH

In a study published in 2011 in *Clinical Nutrition*, researchers from Germany noted that infant formula made with canola oil is widely used in Europe, Australia, Asia, and Latin America. However, because of safety concerns, it is not used in North America. Are these safety concerns based on fact? To learn more about the issue, the researchers compared the growth of full-term infants fed formulas with and without canola oil from week 4 to month 7. The cohort consisted of 85 infants: 60 consumed a formula with canola oil; 25 consumed a formula without canola oil. The researchers found that weight and length measures did not differ between the infants fed formula with canola oil or the formula without canola oil. The researchers wrote that their investigation showed "no significant differences in absolute or standardized weight gain during the first seven months of life among infants fed formula containing canola oil . . . and infants fed formula without canola oil." They concluded that "infant formulas containing part of the lipid source as canola oil have no adverse effect on infant growth, neither in weight nor in length compared to feeding formula without canola oil."[10]

IS CANOLA OIL BENEFICIAL?

Canola oil appears to have a number of different beneficial properties. Still, until more information is learned, people who may be at increased risk for a stroke may wish to limit their intake.

NOTES

1. U.S. Canola Association. www.uscanola.com.
2. CanolaInfo. www.canolainfo.org.
3. Bhatia, E., C. Doddivenaka, X. Zhang et al. 2011. "Chemopreventive Effects of Dietary Canola Oil on Colon Cancer Development." *Nutrition and Cancer* 63(2): 242–247.
4. Wang, J., E. M. John, P. L. Horn-Ross, and S. A. Ingles. 2008. "Dietary Fat, Cooking Fat, and Breast Cancer Risk in a Multiethnic Population." *Nutrition and Cancer* 60(4): 492–504.
5. Ion, G., J. A. Akinsete, and W. E. Hardman. March 6, 2010. "Maternal Consumption of Canola Oil Suppressed Mammary Gland Tumorigenesis in C3(1) Tag Mice Offspring." *BMC Cancer* 10: 81+.

6. Esmaillzadeh, A. and L. Azadbakht. October 2008. "Home Use of Vegetable Oils, Markers of Systemic Inflammation, and Endothelial Dysfunction among Women." *The American Journal of Clinical Nutrition* 88(4): 913–921.

7. Gillingham, L.G., J.A. Gustafson, S.Y. Han et al. February 2011. "High-Oleic Rapeseed (Canola) and Flaxseed Oils Modulate Serum Lipids and Inflammatory Biomarkers in Hypercholesterolaemic Subjects." *British Journal of Nutrition* 105(3): 417–427.

8. Papazzo, A., X.A. Conlan, L. Lexis, and P.A. Lewandowski. 2011. "Differential Effects of Dietary Canola and Soybean Oil Intake on Oxidative Stress in Stroke-Prone Spontaneously Hypertensive Rats." *Lipids in Health and Disease* 10(1): 98–106.

9. Hariri, N., R. Gougeon, and L. Thibault. September 2010. "A Highly Saturated Fat-Rich Diet Is More Obesogenic than Diets with Lower Saturated Fat Content." *Nutrition Research* 30(9): 632–643.

10. Rzehak, P., S. Koletzko, B. Koletzko et al. June 2011. "Growth of Infants Fed Formula Rich in Canola Oil (Low Erucic Acid Rapeseed Oil)." *Clinical Nutrition* 30(3): 339–345.

REFERENCES AND RESOURCES

Magazines, Journals, and Newspapers

Bhatia, E., C. Doddivenaka, X. Zhang et al. 2011. "Chemopreventive Effects of Dietary Canola Oil on Colon Cancer Development." *Nutrition and Cancer* 63(2): 242–247.

Esmaillzadeh, A. and L. Azadbakht. October 2008. "Home Use of Vegetable Oils, Markers of Systemic Inflammation, and Endothelial Dysfunction among Women." *The American Journal of Clinical Nutrition* 88(4): 913–921.

Gillingham, L.G., J.A. Gustafson, S.Y. Han et al. February 2011. "High-Oleic Rapeseed (Canola) and Flaxseed Oils Modulate Serum Lipids and Inflammatory Biomarkers in Hypercholesterolaemic Subjects." *British Journal of Nutrition* 105(3): 417–427.

Hariri, N., R. Gougeon, and L. Thibault. September 2010. "A Highly Saturated Fat-Rich Diet Is More Obesogenic than Diets with Lower Saturated Fat Content." *Nutrition Research* 30(9): 632–643.

Ion, G., J.A. Akinsete, and W.E. Hardman. March 6, 2010. "Maternal Consumption of Canola Oil Suppressed Mammary Gland Tumorigenesis in C3(1) Tag Mice Offspring." *BMC Cancer* 10: 81+.

Papazzo, A., X.A. Conlan, L. Lexis, and P.A. Lewandowski. 2011. "Differential Effects of Dietary Canola and Soybean Oil Intake on Oxidative Stress in Stroke-Prone Spontaneously Hypertensive Rats." *Lipids in Health and Disease* 10(1): 98–106.

Rzehak, P., S. Koletzko, B. Koletzko et al. June 2011. "Growth of Infants Fed Formula Rich in Canola Oil (Low Erucic Acid Rapeseed Oil)." *Clinical Nutrition* 30(3): 339–345.

Wang, J., E.M. John, P.L. Horn-Ross, and S.A. Ingles. 2008. "Dietary Fat, Cooking Fat, and Breast Cancer Risk in a Multiethnic Population." *Nutrition and Cancer* 60(4): 492–504.

Websites

Canola Council of Canada. www.canolacouncil.org.
CanolaInfo. www.canolainfo.org.
U.S. Canola Association. www.uscanola.com.

Coconut Oil: Overview

Until fairly recent times, almost all health researchers and medical providers viewed coconut oil as a food that should be avoided, especially by those at risk for cardiovascular problems. It was generally assumed that coconut oil aggressively clogged arteries and sent cholesterol levels soaring. After all, coconut oil contains large amounts of saturated fat, and it is widely believed that saturated fat harms the body. For decades, medical providers had been telling people to reduce the amount of saturated fat in their diets.

But, those days of shunning coconut oil appear to be ancient history. Today, it is widely praised. In fact, some of the best-known health care and medical advocates now advise including coconut oil in the daily diet. Supposedly, it supports cardiovascular health, is beneficial for people with type 2 diabetes, and is useful for controlling weight. It is thought to make skin and hair shine, and it may well improve memory. In fact, the list of positive medical claims now associated with coconut oil appears endless.

It turns out that all saturated fats are not alike. The saturated fats in coconut oil are medium-chain fatty acids. They have stearic acid and lauric acid, which are known to support health. Moreover, the saturated fats in coconut oil are used directly by the body to produce energy. The saturated fats associated with heart

Coconuts and organic coconut oil. (Geografika/Dreamstime.com)

disease are myristic and palmitic acids, which are in animal products, such as meat and fatty dairy products. A 2006 article in *The Ceylon Medical Journal* noted that "the fatty acid composition of the saturated fats in coconut is very different from that of saturated fats from animal sources."[1]

Coconut oil is moderately priced and is readily available in supermarkets and specialty stores. It is also easy to purchase online.

This entry discusses a wide array of health benefits coconut oil may provide. The next entry looks specifically at coconut oil's influence on cardiovascular health.

APPEARS TO SUPPORT BONE HEALTH

In a study published in 2012 in *Evidence-Based Complementary and Alternative Medicine*, researchers from Malaysia investigated the effects of virgin coconut oil on bones in a postmenopausal rat model. The researchers began with 40 female Wistar rats that they randomly divided into five groups. To induce menopause and to reduce the levels of bone-building estrogen in the blood, the rats in three of the groups were ovariectomized. For six weeks, one group of ovariectomized rats was fed virgin coconut oil supplementation and a second one was fed calcium. One group of ovariectomized rats served as the control. The researchers found that the control rats lost a considerable amount of bone. On the other hand, the ovariectomized rats on virgin coconut oil did not have the bone loss. According to the researchers, the virgin coconut oil appeared "to reverse the effects of estrogen deficiency on the bone structure." The six weeks of virgin coconut oil supplementation "prevented bone loss and maintained bone microarchitecture of estrogen-depleted rats." In fact, the researchers determined that supplementation with virgin coconut oil was "much better than treatment with calcium." The researchers concluded that "virgin coconut oil could offer an interesting approach to prevent accelerated bone loss, especially in postmenopausal women."[2]

MAY HELP REDUCE BELLY FAT

In an open-label pilot study published in 2011 in *ISRN Pharmacology*, researchers from Malaysia wanted to learn if four weeks of coconut oil supplementation would help reduce the accumulation of fat in the belly area. The cohort consisted of 20 obese but otherwise healthy subjects between the ages of 24 and 51. The researchers found that the men in the study experienced significant reductions in their belly fat; similar results were not seen in the women. While the women had reductions in belly fat, they were not statistically significant. The virgin coconut oil did not appear to have "any deleterious effects on biochemical and organ functions."[3]

SEEMS TO COMBAT TOOTH DECAY

In a report presented to the 2012 Annual Conference of the Society of General Microbiology, researchers from the Athlone Institute of Technology in Ireland revealed that coconut oil is able to attack the bacteria that cause tooth decay. According to these researchers, after testing coconut oil against strains of *Streptococcus* bacteria, which is frequently found in the mouth, they learned that coconut oil is a natural antibiotic. As a result, it could be added to commercial dental care products and reduce the incidence of dental decay. The lead researcher, Dr. Damien Brady, noted that "dental caries is a commonly overlooked health problem affecting 60–90% of children and the majority of adults in industrialized countries."

Moreover, "incorporating enzyme-modified coconut oil into dental hygiene products would be an attractive alternative to chemical additives, particularly as it works at relatively low concentrations."[4]

MAY BE USEFUL FOR MASSAGING BABIES

In an open, randomized controlled study published in 2005 in *Indian Pediatrics*, researchers from India compared the effects of massaging newborns with coconut oil, mineral oil, or a placebo. Beginning on their second day of life, the babies were massaged by a trained person until they were discharged. After discharge, the mothers, who had been taught the massage technique, continued the massage until their babies were 31 days old. Every day, the mothers found time for four 5-minute massages. When compared to the control babies, the babies massaged with coconut oil had better weight gain. According to the researchers, by the time the babies were 31 days old, "the weight was significantly higher in the coconut oil subgroup as compared to placebo as well as the mineral oil group."[5]

In a more recent article published in 2010 in *Infant Behavior and Development*, researchers from Miami, Florida; Santa Barbara, California; and Tuscaloosa, Alabama, conducted a review of preterm infant massage therapy studies. They found that the use of oils, including coconut oil and safflower oil, "enhanced the average weight gain." The weight gain was associated with shorter hospital stays and significant cost savings. In addition, "the transcutaneous absorption of oil . . . increased triglycerides."[6]

MAY BE USEFUL FOR COLITIS

In a study published in 2009 in *The Journal of Nutrition*, researchers from Barcelona, Spain, and Zurich, Switzerland, wanted to learn if eight weeks of coconut oil supplementation would be useful for mice bred to have colitis. The researchers began by placing their mice on one of two different diets. The control mice were placed on a diet with sunflower oil as the source of fat; the other mice were placed on a diet in which half of the fat was from coconut oil and

half came from sunflower oil. When compared to the control mice, the incidence of colitis was significantly lower in the mice on coconut and sunflower
oils. The researchers concluded that coconut oil has a "possible primary therapeutic effect . . . in human inflammatory bowel disease."[7]

MAY BE USEFUL FOR ALZHEIMER'S DISEASE

While there is little direct research on the use of coconut oil for Alzheimer's
disease, there is a good deal of information compiled by Mary T. Newport, MD, a
neonatologist, who placed her husband of 36 years on supplemental coconut oil to
treat his early onset Alzheimer's disease. When Steve Newport was first diagnosed,
his physicians treated him with FDA-approved medications. But, his decline continued. Dr. Newport became particularly alarmed when her husband lost interest
in eating and stopped kayaking and gardening, activities he once enjoyed.

So, Dr. Newport started researching to find an alternative approach. Eventually, she learned that certain brain cells of people with Alzheimer's disease have
a problem metabolizing glucose, which is the main source of energy for the brain.
Deprived of the fuel they need to live, the brain cells start dying. But, there was
a way to feed these cells. Ketones. And how does one obtain ketones? When a
person consumes medium-chain triglycerides, such as nonhydrogenated coconut
oil, the liver directly converts them into ketones. (More detailed information on
this process may be found on Dr. Newport's website Coconut Oil and Ketones,
www.coconutketones.com.)

After Steve was told that he was too ill to participate in a clinical study,
Dr. Newport decided to add coconut oil supplementation to his diet. Soon, it
appeared that coconut oil "lifted the fog." After five days of supplementation,
there was "tremendous improvement." According to a 2008 article in the *Tampa
Bay Times*, five months after Steve began the coconut oil supplementation, his
tremors subsided, his visual disturbances disappeared, and he became "more social and interested in those around him."

According to Dr. Newport, it is easy to incorporate coconut oil into the diet of
people with Alzheimer's disease. It may be added to cereal, such as oatmeal, or to
soups, salads dressings, sauces, and smoothies. It may be used on potatoes, pasta,
and rice. When introducing coconut oil as a supplement to the diet, it is best to
begin with a small amount, such as one teaspoon. Dr. Newport advices people to
work up to four to six tablespoons per day; it should be added to regular meals.[8]
That may well be too much for some people. Consuming higher doses of coconut
oil may cause digestive problems, including cramping and diarrhea. Still, smaller
amounts may be helpful to some people with Alzheimer's disease.

MAY BE USEFUL FOR PNEUMONIA IN CHILDREN

According to research presented to the 2008 annual meeting of the American
College of Chest Physicians and published in their journal *CHEST*, also in 2008,

researchers from the Philippines randomly assigned 40 pediatric patients with community-acquired pneumonia to receive intravenous ampicillin plus oral virgin coconut oil or intravenous ampicillin alone. The virgin coconut oil was given for three days in a row. The researchers found that the respiratory rate of the children on virgin coconut oil normalized faster than the children not on the supplement—32.6 hours versus 48.2 hours. Moreover, after 72 hours, the children on the virgin coconut oil had fewer crackles (added breath sounds in lungs)—25% versus 60%. In addition, the children taking virgin coconut oil had a quicker normalization of temperature and oxygen saturation, and they were hospitalized for fewer hours—75.9 hours versus 91.85 hours. The researchers concluded that "virgin coconut oil is an effective adjunct therapy for pediatric community acquired pneumonia in accelerating the normalization of respiratory rate and resolution of crackles."[9]

MAY BE USEFUL FOR HYPOGLYCEMIC
EPISODES IN PEOPLE WITH TYPE 1 DIABETES

In a trial published in 2009 in *Diabetes,* researchers from New Haven, Connecticut, and New York attempted to determine if the ingestion of medium-chain triglycerides, such as those found in coconut oil, could improve the impaired cognition associated with hypoglycemia in people with type 1 diabetes. During two separate acute-insulin-induced hypoglycemia sessions, 11 people with type 1 diabetes randomly received either medium-chain triglycerides or placebo drinks, while a battery of cognitive tests were performed. The researchers found that the "ingestion of medium-chain triglycerides reversed these effects." In fact, the researchers noted the medium-chain triglycerides proved to be a rapidly absorbed fuel that improved cognition during periods of low blood sugar. And, the medium-chain triglycerides did this without appearing to cause any additional harm to the body.[10]

SEEMS TO BE EFFECTIVE AGAINST
ADULT ATOPIC DERMATITIS

In a double-blind controlled trial published in 2008 in *Dermatitis,* researchers from the Philippines treated 52 adults who had atopic dermatitis, an itchy inflammation of the skin that is often colonized with *Staphylococcus aureus,* with either virgin coconut oil or virgin olive oil (control). Of the 26 subjects on virgin coconut oil, 20 tested positive for *Staphylococcus aureus* at baseline; of the 26 subjects on virgin olive oil, 12 tested positive at baseline. After treatment with virgin coconut oil, only one subject remained positive; after treatment with virgin olive oil, six subjects remained positive. The coconut oil was more effective than the olive oil. And, that did not surprise the researchers. They noted that virgin coconut oil "has a long tradition of use in treating infections." And, they concluded that "a history of safe topical use and no known or reported cases of contact dermatitis, along with its duel effects as moisturizer and antiseptic, opens up more research and clinical possibilities for virgin coconut oil."[11]

HAS ANTI-INFLAMMATION, ANTIPAIN, AND ANTIFEVER PROPERTIES

In a study published in 2010 in *Pharmaceutical Biology*, researchers from Thailand examined the anti-inflammation, antipain, and antifever properties of virgin coconut oil in male Sprague-Dawley rats and male Swiss albino mice. During the course of a number of different laboratory tests, the researchers found that virgin coconut oil demonstrated these properties. For example, virgin coconut oil "caused significant reductions of ear edema, paw edema, and granuloma [small nodule] formation." When used in higher doses, virgin coconut oil had anti-inflammatory properties "on both acute and chronic phases of inflammation." Moreover, an acetic acid–induced writhing response test showed the antipain properties and a yeast-induced hyperthermia test found antifever properties.[12]

USEFUL AS AN ANTIMICROBIAL AGENT

In a study published in 2005 in *Molecular and Cellular Biochemistry*, researchers from Washington, D.C. and Bethesda, Maryland commented that the world is increasingly seeing the emergence of numerous antimicrobial-resistant organisms. In addition, large numbers of commercially manufactured antimicrobial agents may have serious side effects, which has the potential to limit their long-term and even short-term use. "Hence, the accepted practice is to encourage the use of antimicrobial agents only when necessary to treat infections, thus precluding their prophylactic use under many circumstances."

That is why it is so important to find more natural products that have antimicrobial properties. "Natural products, many of which can be used for long periods, might be less likely to produce side effects, and resistance . . . has not been shown." As a result, these researchers tested monolaurin (once in the body, lauric acid turns into monolaurin) against several types gram-positive and gram-negative bacteria. And, they found that monolaurin was useful in the prevention and treatment of some bacterial infections, especially those that are difficult to heal or antibiotic resistant. These results were obtained both in petri dishes and in mice. The researchers found that monolaurin was indeed effective against bacteria. For example, monolaurin was "exquisitely effective against *H. pylori*, a gram-negative organism that is difficult to culture." Why is that significant? "Because approximately two-thirds of the world's human population is colorized or infected with this organism, a safe and effective herb-derived and/or natural fat therapy would provide an alternative to the current antimicrobial therapy." The researchers concluded that monolaurin is an effective antimicrobial agent that is useful against bacterial infections. And, because natural substances like monolaurin have proven safety records, they may be used for longer periods of time and even for the prevention of bacterial infections.[13]

MAY BE USEFUL FOR BENIGN
PROSTATIC HYPERPLASIA

In a study published in 2007 in the *Journal of Pharmacy and Pharmacology*, researchers from Cuba wanted to learn if coconut oil would be helpful for men with benign prostatic hyperplasia, the noncancerous enlargement of the prostate gland that may lead to problems with urination, frequently seen in men as they age. The cohort consisted of seven groups of 10 Sprague-Dawley rats. The rats in six of the groups were injected with testosterone to induce prostate enlargement. After the rats in two of the groups served as the controls, the rats in five of the groups were treated with a lower and high dose of coconut oil, one dose of saw palmetto lipid extract (which is believed to be helpful for prostate enlargement), or a lower and higher dose of sunflower oil. After 14 days of treatment, the rats were sacrificed and the prostate organs were removed and weighed. The researchers found that both doses of coconut oil significantly reduced both the weight of the prostate glands and the ratio of prostate weight to body weight. (They saw palmetto lipid extract and the higher dose of sunflower oil also reduced the prostate weights and the ratios.)[14]

USEFUL FOR TREATING ACUTE
ALUMINUM PHOSPHIDE POISONING

In a study published in 2005 in *Human & Experimental Toxicology*, researchers from Iran explained that aluminum phosphide is used in grain storage facilities to control rodents and pests. It produces phosphine gas, which is poisonous. Though the researchers noted that there is no known antidote for phosphine gas, they wondered if coconut oil could help prevent its absorption into the body. They tested their hypothesis in a 28-year-old man who had ingested a lethal amount of aluminum phosphide, about six hours before his hospital admission, in order to kill himself. According to the researchers, he "had signs and symptoms of severe toxicity." In addition to the usual treatments for oral poisoning, such as gastric lavage and the oral administration of activated charcoal, the patient consumed coconut oil. "All of these factors, especially coconut oil, were probably important in preventing what would otherwise have been a fatal ingestion of aluminum phosphide."[15]

MAY BE USEFUL FOR PEOPLE WITH
TYPE 2 DIABETES

In a study published in 2011 in the *Journal of Diabetes Mellitus*, researchers from India wanted to learn if cold and hot extracted virgin coconut oil would be useful for people with type 2 diabetes, a medical problem that is strongly associated with obesity and the lack of physical activity. They termed diabetes as "a silent killer" that is "spreading like an epidemic disease." Though there are medications to treat diabetes, the researchers noted that when antidiabetic medications are used for a longer period of time, they may have "undesirable effects." So, it is important for researchers to find "natural anti-diabetic foods with [the] least ill effects and low cost" to manage type 2 diabetes.

The researchers began by dividing 32 male Wistar albino rats into four groups of 8 rats. The rats in the first group were nondiabetic controls; the rats in the three remaining groups were injected with a substance to induce diabetes. The rats in the second group were then fed a commercial coconut oil; the rats in the third groups were fed cold extracted virgin coconut oil; and the rats in the fourth group were fed hot extracted virgin coconut oil. The trial continued for 21 days. After monitoring blood sugar levels, the researchers found that the greatest reductions in blood glucose were seen in the rats fed the hot extracted virgin coconut oil. The hot extracted virgin coconut oil also had greater free radical scavenging activity than the other two oils. The researches concluded that "the better health benefits of HEVCO [heat extracted virgin coconut oil] may be attributed to its higher PP [polyphenolic] content and also possible increased bioavailability of nutrients."[16]

In a study published in 2009 in *Diabetes*, researchers from Australia and China wanted to determine the effect that medium-chain fatty acids, such as those found in coconut oil, had on the action of insulin in the bodies of mice and male Wistar rats. During a five-week trial, the diets of the mice and rats were supplemented with either a low fat diet or a high fat diet containing lard or a diet with medium-chain fatty acids from hydrogenated coconut oil. The researchers found that the coconut oil provided protection against insulin resistance—an impaired ability of the body's muscle and fat cells to respond to insulin. Coconut oil also prevented the accumulation of body fat that would normally be seen in other high fat diets with similar caloric content. The researchers noted that their findings "clearly show that MCFAs [medium-chain fatty acids] do not induce insulin resistance in either muscle or adipose tissue."[17]

USEFUL FOR THE DIETARY TREATMENT OF DRUG-RESISTANT PEDIATRIC EPILEPSY

In an article published in 2013 in the *Biomedical Journal*, a researcher from Canada and a second researcher from Taiwan argued for the increased use of a ketogenic diet, containing a percentage of medium-chain triglycerides, for drug-resistant pediatric epilepsy. According to these researchers, too much of the literature on ketogenic diets has focused on the potential gastrointestinal side effects of this type of diet, such as diarrhea, vomiting, bloating, and cramps. As a result, the diet "has been an underutilized diet therapy for intractable epilepsy among children."

The researchers noted that children who are placed on a medium-chain tri-glyceride ketogenic diet are generally admitted to the hospital. "The meals are typically advanced daily in one-third caloric intervals of 6 feeds until full calorie meals are tolerated, while keeping the percentage of the MCT oil consistent." When the children experience diarrhea or vomiting, the percentage of oil is low-ered by 10% at the next meal. Once the children are able to tolerate this high fat diet, they are discharged. After that, there are follow-up communications and periodic visits with dieticians. Children should be carefully monitored to en-sure they are receiving adequate nutrition for optimum growth and development. "During the fine-tuning stage, the dietician monitors the patient's progress and modifies the diet to achieve improved seizure control. Nutritional assessments and nutrients adjustments are also essential to assure best growth and optimal nutritional status for each individual." The need for vitamin supplementation with this diet is not uncommon. (It is important to note that it is a good idea to avoid this diet in children with liver problems. It has the potential to worsen the condition.)

The researchers noted that when this diet is effective, it may be useful for as many as 6–12 years. And, when it is finally discontinued, it is best to taper the process over a period of time. The researchers concluded their article by advocat-ing for the increased use of this diet for children with drug-resistant epilepsy.[18]

MAY BE USEFUL FOR NICOTINE
DEPENDENCE AND RELAPSE

In a study published in 2011 in the *International Journal of Pharmacology*, re-searchers from Indonesia wanted to learn if virgin coconut oil would be useful for mitigating the addictive properties associated with nicotine, the psychoactive ingredient found in tobacco, and the frequency of relapse among smokers. The researchers assigned rats to different groups before exposing them to nicotine treatments. Some of the rats were given virgin coconut oil before these exposures. The researchers found that the virgin coconut oil provided a degree of protec-tion from nicotine dependence. "Pretreatment with VCO [virgin coconut oil] prevented the development of nicotine dependence, as shown by the decrease in preference score in rats receiving VCO prior to every nicotine dose."

An interesting result was observed when nicotine was reintroduced to absti-nent rats that had previously been exposed to nicotine. "VCO also prevented the increase in preference score in abstinent rats challenged with nicotine in-jection, indicating a blocking effect of VCO on relapse to nicotine." This has an important clinical application. For in the clinical setting, relapse "has been a main hurdle in the treatment of drug addiction." In fact, "addicts who have been drug-free, even for years, may still resume the drug taking habit when they are exposed to the drug or drug cues." According to these researchers, virgin coconut oil has the potential to play an important role in the prevention and treatment of nicotine addiction.[19]

MAY HAVE ANTIFUNGAL PROPERTIES FOR VAGINAL CANDIDIASIS

In a study published in 2008 in *Microbiology Indonesia,* researchers from Indonesia wanted to learn if virgin coconut oil would be useful for vaginal candidiasis, a common vaginal fungal infection that causes a number of uncomfortable symptoms such as itching, burning, and pain. The cohort, which consisted of 30 women with vaginal candidiasis, was divided into three groups of 10. For two months, the women in one group were treated with two tablespoons per day of zinc-enriched virgin coconut oil, the women in a second group were treated with one tablespoon per day of zinc-enriched virgin coconut oil, and the women in a third group were given placebos. Vaginal secretions were taken at three different times—at baseline, at one month, and after two months. After two months, the women in both treatment groups experienced similar reductions of fungal colonies, but the number of colonies was still higher than normal.[20]

MAY HAVE LIVER PROTECTION PROPERTIES

In a study published in 2011 in *Evidence-Based Complementary and Alternative Medicine,* researchers from Malaysia noted that liver diseases are common medical problems that are seen throughout the world. At the same time, "conventional drugs used in the treatment of liver disease are sometimes inadequate and can have serious adverse effects." So, they wondered if virgin coconut oil might help protect the liver. The cohort consisted of nine different groups of rats. After setting aside some rats to serve as controls, rats were pretreated with two different types of coconut oil. Then, the researchers used paracetamol, a known liver toxin, to induce liver toxicity in the rats. They found that the highest concentrations of both types of virgin coconut oil provided a degree of liver protection. The researchers concluded that virgin coconut oil "possesses a promising hepatoprotective effect."[21]

WOUND-HEALING PROPERTIES

In a study published in 2008 in the *Indian Journal of Pharmacology,* researchers from Manipal, India, compared the effectiveness of coconut oil and coconut oil combined with silver sulphadiazine, an antibacterial cream used for burns, on burn wounds on rats. The researchers began by dividing rats with burn wounds into four groups of six. The rats in the control group received no treatment for their wounds, the rats in the second group were treated with silver sulphadiazine cream, the rats in the third group were treated with coconut oil, and the rats in the fourth group were treated with silver sulphadiazine and coconut oil. The researchers found "significant improvement in burn wound contraction in the rats treated" with coconut oil or coconut oil combined with silver sulphadiazine. The researchers conclude that coconut oil "could significantly enrich the assortment of topical medications available for the treatment of burns."[22]

In a study published in 2010 in *Skin Pharmacology and Physiology*, different researchers from India also examined the healing properties of two different concentrations of virgin coconut oil on rat wounds. The researchers divided their rats into three groups of six. After the wounds were created, one group of rats received no treatment. That group consisted of the control rats. The wounds of the rats in the second and third groups were treated with virgin coconut oil. After 10 days of treatment, the wounds were evaluated. The researchers found that the wounds treated with virgin coconut oil "healed much faster." The virgin coconut oil increased the speed in which the skin repaired itself. The researchers concluded that virgin coconut oil "beneficially affects the wound environment and fastens the healing process."[23]

MAY BE USEFUL FOR ACNE

In a study published in 2009 in the *Journal of Investigative Dermatology*, researchers from San Diego, California, Germany, and Taiwan investigated the in vitro and in vivo use of lauric acid, which is found in abundance in coconut oil, to kill the skin bacteria associated with acne. These bacteria include *Propionibacterium acnes*, *S. aureus*, and *Staphylococcus epidermidis*. In both their in vitro and in vivo (mouse ear) studies, the researchers found that lauric acid had a potent ability to prevent the growth of skin bacteria. For example, the researchers learned that lauric acid was 15 times more potent in killing *P. acnes* than benzoyl peroxide, a medication commonly used to treat acne. It was also more effective than benzoyl peroxide in reducing swelling and inflammation. And, lauric acid had no adverse effects; benzoyl peroxide is known to cause irritation, burning, skin peeling, rashes, redness, and stinging. The researchers concluded that their work "demonstrated the antimicrobial property of lauric acid against *P. acnes in vitro* and its therapeutic effects on *P. acnes*-induced inflammation *in vivo* using the ICR mouse ear model." In addition, lauric acid was proven to be "an alternative treatment option to the antibiotic therapy of acne vulgaris."[24]

Three of the researchers in the previous study joined with other San Diego area researchers for a follow-up study that was published in 2009 in *Biomaterials*. They wanted to learn if encapsulating lauric acid in liposomes (small sacs) would enhance the ability of lauric acid to kill bacteria and heal acne. The researchers found that the liposomes filled with lauric acid adhered to the *P. acnes* bacteria; the lauric acids entered the bacterial membrane and effectively killed the bacteria.

One problem with this system is the tendency of liposomes to clump together. To prevent this from happening, the researchers attached nanoparticles, also known as nanobombs, to the surface of the lauric acid–filled liposomes. Once the liposomes attach to the bacterial membrane, the nanoparticles fall off. Then the lauric acid is able to kill the bacteria. The researchers noted that "since LA [lactic acid] is a natural compound that is the main acid in coconut oil," coconut oil "holds great potential of becoming an innate, safe and effective therapeutic medication for acne vulgaris and other *P. acnes* associated diseases."[25]

USEFUL FOR KILLING LICE

In a randomized, controlled, parallel group trial published in 2010 in the *European Journal of Pediatrics*, researchers from the United Kingdom compared the ability of permethrin, a synthetic chemical, and a combination of coconut and anise oil spray to destroy an infestation of lice. The cohort consisted of 100 people with active head lice infestations. Half were treated on two separate occasions with permethrin; half were treated also on two separate occasions with the coconut and anise oil spray. Ninety-six of the participants completed the trial. After the first treatments of both permethrin and the coconut and anise oil spray, some lice remained. So, a second treatment was always necessary. The researchers found that about one-third of their subjects who received treatment with permethrin failed to improve. On the other hand, most of the people who were treated with the coconut and anise oil spray improved. The researchers noted that they believe "coconut and anise spray should remain a viable treatment option for most people well into the future."[26]

MAY BE USEFUL AGAINST THE PARASITIC DISEASE TUNGIASIS (SAND FLEA DISEASE)

In an article published in 2013 in the *European Journal of Clinical Microbiology & Infectious Diseases*, researchers from Berlin, Germany, Kampala, Uganda, and Skövde, Sweden, discussed the serious parasitic illness known as sand flea disease and the use of coconut oil to prevent it. According to the researchers, the parasites enter the body through the skin of the feet. It causes acute inflammation, edema, itching, deep ulcers, severe clinical pathology, deformation of toes, loss of nails, and intense pain. There are anecdotal reports of people cutting off their toes to reduce their extreme pain. Moreover, in several countries in sub-Sahara Africa, it is affecting people in epidemic numbers. And, it has become "the most common parasitic skin disease acquired by tourists frequenting tropical beaches."

Yet, according to the researchers, there is a way to prevent this illness and its associated symptoms. "Randomized controlled trials have shown that the regular application of a repellent with coconut oil reduces the attack rate by 92% to almost 100% and prevents the development of morbidity."[27]

MAY BE USEFUL FOR CHILDHOOD DIARRHEA

In a double-blind randomized study published in 2007 in the *Asian Pacific Journal of Clinical Nutrition*, researchers from the Philippines wanted to learn if a medium-chain triglyceride oil, such as coconut oil, would be useful for reducing the incidence and duration of childhood diarrhea. The cohort consisted of 17 children between the ages of 6 and 47 months (mean age of 19.6 months) who had acute diarrhea. According to the researchers, "diarrhea is a major cause of malnutrition owing to low food intake during the illness, reduced nutrient absorption in the intestine, and increased nutrient requirements as a result of infection."

During the study, which continued for one year, eight (47.1%) of the children with diarrhea were treated with a medium-chain triglyceride oil that was added to milk formula or meals and nine (52.9%) were not treated. One subject from the group receiving medium-chain triglyceride developed additional medical problems and had to be dropped from the study.

The researchers found that the children who received the medium-chain triglyceride oil gained weight, while those who did not take the oil lost weight. "A minimal weight gain of about 0.2kg was observed among the MCT [medium-chain triglyceride] group while a weight loss of .03kg was observed among the non-MCT group." The researchers concluded that "using MCT oil in childhood diarrhea may allow weight gain and shorten the required duration of intervention. MCT oil did not cause vomiting, dehydration or fat intolerance. MCT oil did not elevate serum lipid concentrations."[28]

IS COCONUT OIL BENEFICIAL?

Since the research on some of these medical problems is still somewhat limited, it is impossible to know for sure, but it appears that coconut oil does indeed have many health benefits. Because it is so reasonably priced and appears to have so many benefits, many people may want to consider adding coconut oil to their diets. This is especially true for the millions of people dealing with Alzheimer's disease and their caregivers. If coconut oil really has the potential to ease the debilitating symptoms of Alzheimer's disease, it is definitely worth a discussion with a medical provider.

NOTES

1. Amarasiri, W. A. and A. S. Dissanayake. June 2006. "Coconut Fats." *The Ceylon Medical Journal* 51(2): 47–51.

2. Hayatullina, Zil, Norliza Muhammad, Norazlina Mohamed, and Ima-Nirwana Soelaiman. 2012. "Virgin Coconut Oil Supplementation Prevents Bone Loss in Osteoporosis Rat Model." *Evidence-Based Complementary and Alternative Medicine* Article ID 237236: 8 pages.

3. Liau, K. M., Y. Y. Lee, C. K. Chen, and A. H. Rasool. 2011. "An Open-Label Pilot Study to Assess the Efficacy and Safety of Virgin Coconut Oil in Reducing Visceral Adiposity." *ISRN Pharmacology* Article ID 949686: 7 pages.

4. Society for General Microbiology Website. www.sgm.ac.uk.

5. Sankaranarayanan, K., J. A. Mondkar, M. M. Chauhan et al. 2005. "Oil Massage in Neonates: An Open Randomized Controlled Study of Coconut versus Mineral Oil." *Indian Pediatrics* 42(9): 877–884.

6. Field, T., M. Diego, and M. Hernandez-Reif. April 2010. "Preterm Infant Massage Therapy Research: A Review." *Infant Behavior and Development* 33(2): 115–124.

7. Mañé, Josep, Elisabet Pedrosa, Violeta Lorén et al. March 2009. "Partial Replacement of Dietary (n-6) Fatty Acids with Medium-Chain Triglycerides Decreases the Incidence of Spontaneous Colitis in Interleukin-10-Deficient Mice." *The Journal of Nutrition* 139(3): 603–610.

8. Coconut Oil and Ketones Website. www.coconutketones.com.

9. Erguiza, Gilda Sapphire, Arnel Gerald Jiao, Michelle Reley, and Shelesh Ragaza. October 2008. "The Effect of Virgin Coconut Oil Supplementation for Community-Acquired Pneumonia in Children Aged 3 to 60 Months Admitted at the Philippine Children's Medical Center: A Single Blinded Randomized Controlled Trial." *CHEST* 134(4): Meeting Abstracts.

10. Page, K. A., A. Williamson, N. Yu et al. May 2009. "Medium-Chain Fatty Acids Improve Cognitive Function in Intensively Treated Type 1 Diabetic Patients and Support *In Vitro* Synaptic Transmission during Acute Hypoglycemia." *Diabetes* 58(5): 1237–1244.

11. Verallo-Rowell, V. M., K. M. Dillague, and B. S. Syah-Tjundawan. November–December 2008. "Novel Antibacterial and Emollient Effects of Coconut and Virgin Olive Oils in Adult Atopic Dermatitis." *Dermatitis* 19(6): 308–315.

12. Intahphuak, S., P. Khonsung, and A. Panthong. 2010. "Anti-Inflammatory, Analgesic, and Antipyretic Activities of Virgin Coconut Oil." *Pharmaceutical Biology* 48(2): 151–157.

13. Preuss, Harry G., Bobby Echard, Mary Enig et al. 2005. "Minimum Inhibitory Concentrations of Herbal Essential Oils and Monolaurin for Gram-Positive and Gram-Negative Bacteria." *Molecular and Cellular Biochemistry* 272(1–2): 29–34.

14. Arruzazabala, Mar?a de Lourdes, Vivian Molina, Rosa Más et al. 2007. "Effects of Coconut Oil on Testosterone-Induced Prostatic Hyperplasia in Sprague-Dawley Rats." *Journal of Pharmacy and Pharmacology* 59(7): 995–999.

15. Shadnia, Shahin, Mojgan Rahimi, Abdolkarim Pajoumand et al. 2005. "Successful Treatment of Acute Aluminum Phosphide Poisoning: Possible Benefit of Coconut Oil." *Human & Experimental Toxicology* 24(4): 215–218.

16. Siddalingaswamy, Mahadevappa, Arunchand Rayaorth, and Farhath Khanum. 2011. "Anti-Diabetic Effects of Cold and Hot Extracted Virgin Coconut Oil." *Journal of Diabetes Mellitus* 1(4): 118–123.

17. Turner, Nigel, Krit Hariharan, Jennifer TidAng et al. November 2009. "Enhancement of Muscle Mitochondrial Oxidative Capacity and Alterations in Insulin Action Are Lipid Species Dependent." *Diabetes* 58(11): 2547–2554.

18. Liu, Y. M. and H. S. Wang. January–February 2013. "Medium-Chain Triglyceride Ketogenic Diet, an Effective Treatment for Drug-Resistant Epilepsy and a Comparison with Other Ketogenic Diets." *Biomedical Journal* 36(1): 9–15.

19. Anggadiredja, Kusnandar, Anggraini Barlian, Yangie Dwi Marga Pinang, and Dian Anggraeny. 2011. "Virgin Coconut Oil Prevents Nicotine Dependence and Relapse." *International Journal of Pharmacology* 7(5): 664–669.

20. Winarsi, Hery, Hernayanti, and Agus Purwanto. August 2008. "A Supplement Based on Zn-Enriched Virgin Coconut Oil as an Antifungal Agent for Vaginal Candidiasis Patients." *Microbiology Indonesia* 2(2): 69–72.

21. Zakaria, Z. A., M. S. Rofiee, M. N. Somchit et al. 2011. "Hepatoprotective Activity of Dried- and Fermented-Processed Virgin Coconut Oil." *Evidence-Based Complementary and Alternative Medicine* Article ID 142739: 8 pages.

22. Srivastava, P. and S. Durgaprasad. July–August 2008. "Burn Wound Healing Property of *Cocos nucifera*: An Appraisal." *Indian Journal of Pharmacology* 40(4): 144–146.

23. Nevin, K. G. and T. Rajamohan. September 2010. "Effect of Topical Application of Virgin Coconut Oil on Skin Components and Antioxidant Status during Dermal Wound Healing in Young Rats." *Skin Pharmacology and Physiology* 23(6): 290–297.

24. Nakatsuji, Teruaki, Mandy C. Kao, Jia-You Fang et al. October 2009. "Antimicrobial Property of Lauric Acid against *Propionibacterium acnes*: Its Therapeutic Potential for Inflammatory Acne Vulgaris." *Journal of Investigative Dermatology* 129(10): 2480–2488.

25. Yang, Darren, Dissaya Pornpattananangkul, Teruaki Nakatsuji et al. 2009. "The Antimicrobial Activity of Liposomal Lauric Acids against *Propionibacterium acnes*." *Biomaterials* 30(30): 6035–6040.

26. Burgess, Ian F., Elizabeth R. Brunton, and Nazma A. Burgess. January 2010. "Clinical Trial Showing Superiority of a Coconut and Anise Spray over Permethrin 0.43% Lotion for Head Louse Infestation, ISRCTN96469780." *European Journal of Pediatrics* 169(1): 55–62.

27. Feldmeier, H., E. Sentongo, and I. Krantz. January 2013. "Tungiasis (Sand Flea Disease): A Parasitic Disease with Particular Challenges for Public Health." *European Journal of Clinical Microbiology & Infectious Diseases* 32(1): 19–26.

28. Tanchoco, Celeste C., Arsenia J. Cruz, Jossie M. Rogaccion et al. 2007. "Diet Supplemented with MCT Oil in the Management of Childhood Diarrhea." *Asian Pacific Journal of Clinical Nutrition* 16(2): 286–292.

REFERENCES AND RESOURCES

Magazines, Journals, and Newspapers

Amarasiri, W. A. and A. S. Dissanayake. June 2006. "Coconut Fats." *The Ceylon Medical Journal* 51(2): 47–51.

Anggadiredja, Kusnandar, Anggraini Barlian, Yangie Dwi Marga Pinang, and Dian Anggraeny. 2011. "Virgin Coconut Oil Prevents Nicotine Dependence and Relapse." *International Journal of Pharmacology* 7(5): 664–669.

Arruzazabala, Mar?a de Lourdes, Vivian Molina, Rosa Más et al. July 2007. "Effects of Coconut Oil on Testosterone-Induced Prostatic Hyperplasia in Sprague-Dawley Rats." *Journal of Pharmacy and Pharmacology* 59(7): 995–999.

Burgess, Ian F., Elizabeth R. Brunton, and Nazma A. Burgess. January 2010. "Clinical Trial Showing Superiority of a Coconut and Anise Spray over Permethrin 0.43% Lotion for Head Louse Infestation, ISRCTN96469780." *European Journal of Pediatrics* 169(1): 55–62.

Erguiza, Gilda Sapphire, Arnel Gerald Jiao, Michelle Reley, and Shelesh Ragaza. October 2008. "The Effect of Virgin Coconut Oil Supplementation for Community-Acquired Pneumonia in Children Aged 3 to 60 Months Admitted at the Philippine Children's Medical Center: A Single Blinded Randomized Controlled Trial." *CHEST* 134(4): Meeting Abstracts.

Feldmeier, H., E. Sentongo, and I. Krantz. January 2013. "Tungiasis (Sand Flea Disease): A Parasitic Disease with Particular Challenges for Public Health." *European Journal of Clinical Microbiology & Infectious Diseases* 32(1): 19–26.

Field, T., M. Diego, and M. Hernandez-Reif. April 2010. "Preterm Infant Massage Therapy Research: A Review." *Infant Behavior and Development* 33(2): 115–124.

Hayatullina, Zil, Norliza Muhammad, Norazlina Mohamed, and Ima-Nirwana Soelaiman. 2012. "Virgin Coconut Oil Supplementation Prevents Bone Loss in Osteoporosis Rat Model." *Evidence-Based Complementary and Alternative Medicine* Article ID 237236: 8 pages.

Hosley-Moore, Eve. October 28, 2008. "Doctor Says an Oil Lessened Alzheimer's Effects on Her Husband." *Tampa Bay Times*.

Intahphuak, S., P. Khonsung, and A. Panthong. 2010. "Anti-Inflammatory, Analgesic, and Antipyretic Activities of Virgin Coconut Oil." *Pharmaceutical Biology* 48(2): 151–157.

Liau, K.M., Y.Y. Lee, C.K. Chen, and A.H. Rasool. 2011. "An Open-Label Pilot Study to Assess the Efficacy and Safety of Virgin Coconut Oil in Reducing Visceral Adiposity." *ISRN Pharmacology* Article ID 949686: 7 pages.

Liu, Y.M. and H.S. Wang. January–February 2013. "Medium-Chain Triglyceride Ketogenic Diet, an Effective Treatment for Drug-Resistant Epilepsy and a Comparison with Other Ketogenic Diets." *Biomedical Journal* 36(1): 9–15.

Mañé, Josep, Ellisabet Pedrosa, Violeta Lorén et al. March 2009. "Partial Replacement of Dietary (n-6) Fatty Acids with Medium-Chain Triglycerides Decreases the Incidence of Spontaneous Colitis in Interleukin-10-Deficent Mice." *The Journal of Nutrition* 139(3): 603–610.

Nakatsuji, Teruaki, Many C. Kao, Jia-You Fang et al. October 2009. "Antimicrobial Property of Lauric Acid against *Propionibacterium acnes*: Its Therapeutic Potential for Inflammatory Acne Vulgaris." *Journal of Investigative Dermatology* 129(10): 2480–2488.

Nevin, K.G. and T. Rajamohan. September 2010. "Effect of Topical Application of Virgin Coconut Oil on Skin Components and Antioxidant Status during Dermal Wound Healing in Young Rats." *Skin Pharmacology and Physiology* 23(6): 290–297.

Page, K.A., A. Williamson, N. Yu et al. May 2009. "Medium-Chain Fatty Acids Improve Cognitive Function in Intensively Treated Type 1 Diabetic Patients and Support *In Vitro* Synaptic Transmission during Acute Hypoglycemia." *Diabetes* 58(5): 1237–1244.

Preuss, Harry G., Bobby Echard, Mary Enig et al. 2005. "Minimum Inhibitory Concentrations of Herbal Essential Oils and Monolaurin for Gram-Positive and Gram-Negative Bacteria." *Molecular and Cellular Biochemistry* 272(1–2): 29–34.

Sankaranarayanan, K., J.A. Mondkar, M.M. Chauhan et al. 2005. "Oil Massage in Neonates: An Open Randomized Controlled Study of Coconut versus Mineral Oil." *Indian Pediatrics* 42(9): 877–884.

Shadnia, Shahin, Mojgan Rahimi, Abdolkarim Pajoumand et al. 2005. "Successful Treatment of Acute Aluminum Phosphide Poisoning: Possible Benefit of Coconut Oil." *Human & Experimental Toxicology* 24(4): 215–218.

Siddalingaswamy, Mahadevappa, Arunchand Rayaorth, and Farhath Khanum. 2011. "Anti-Diabetic Effects of Cold and Hot Extracted Virgin Coconut Oil." *Journal of Diabetes Mellitus* 1(4): 118–123.

Srivastava, P. and S. Durgaprasad. July–August 2008. "Burn Wound Healing Property of *Cocos nucifera*: An Appraisal." *Indian Journal of Pharmacology* 40(4): 144–146.

Tanchoco, Celeste C., Arsenia J. Cruz, Jossie M. Rogaccion et al. 2007. "Diet Supplemented with MCT Oil in the Management of Childhood Diarrhea." *Asian Pacific Journal of Clinical Nutrition* 16(2): 286–292.

Turner, Nigel, Krit Hariharan, Jennifer TidAng et al. November 2009. "Enhancement of Muscle Mitochondrial Oxidative Capacity and Alterations in Insulin Action Are Lipid Species Dependent." *Diabetes* 58(11): 2547–2554.

Verallo-Rowell, V.M., K.M. Dillague, and B.S. Syah-Tjundawan. November–December 2008. "Novel Antibacterial and Emollient Effects of Coconut and Virgin Olive Oils in Adult Atopic Dermatitis." *Dermatitis* 19(6): 308–315.

Winarsi, Hery, Hernayanti, and Agus Purwanto. August 2008. "A Supplement Based on Zn-Enriched Virgin Coconut Oil as an Antifungal Agent for Vaginal Candidiasis Patients." *Microbiology Indonesia* 2(2): 69–72.

Yang, Darren, Dissaya Pornpattananangkul, Teruaki Nakatsuji et al. 2009. "The Antimicrobial Activity of Liposomal Lauric Acids against *Propionibacterium acnes*." *Biomaterials* 30(30): 6035–6040.

Zakaria, Z.A., M.S. Rofiee, M.N. Somchit et al. 2011. "Hepatoprotective Activity of Dried- and Fermented-Processed Virgin Coconut Oil." *Evidence-Based Complementary and Alternative Medicine* Article ID 142739: 8 pages.

Websites

American Academy of Dermatology. www.aad.org.
Children's National Medical Center. www.childrensnational.org.
Coconut Oil and Ketones. www.coconutketones.com.
Coconut Research Center. www.coconutresearchcenter.org.
Johns Hopkins Medicine. www.hopkinsmedicine.org.
Society for General Microbiology. www.sgm.ac.uk.
Tampa Bay Times. www.tampabaytimes.com.

Coconut Oil and Cardiovascular Health

The previous entry examined coconut oil's many supposed health benefits. This entry discusses studies related to coconut oil's impact on cardiovascular health.

SUPPORTS CARDIOVASCULAR HEALTH

In a study published in 2012 in the *Indian Journal of Experimental Biology*, researchers from India divided male Sprague-Dawley rats into four groups. Rats in the first group took supplemental copra oil (oil made from dried coconut flesh), rats in the second group took supplemental virgin coconut oil, rats in the third group took supplemental olive oil, and rats in the fourth group took supplemental sun-flower oil. After 45 days, the rats were sacrificed, and their cholesterol and tri-glyceride levels were analyzed. When compared to the other three oils, virgin coconut oil was found to support cardiovascular health. "Results indicate that VCO [virgin coconut oil] feeding leads to much lower concentrations of choles-terol and triglycerides in serum and tissues (liver, heart and aorta) and an increase in HDL [high density lipoprotein or "good" cholesterol] cholesterol levels."[1]

In a similar study published in 2011 in *Food and Nutrition Sciences*, research-ers from Malaysia divided 66 Sprague-Dawley male rats into 11 groups. The rats in one group were only fed normal rat pellets; they served as the controls. The rats in the other groups were fed supplemental red palm oil, palm oil, corn oil, or coconut oil for four and eight weeks. After four weeks, the different oils did not appear to affect the total cholesterol. However, after eight weeks, when compared to the control group, the researchers found that the rats in all the treated groups had significant reductions in levels of total cholesterol. And, while eight weeks of treatment with the three other oils increased triglyceride levels, that was not the case for coconut oil. On the other hand, after eight weeks, the other three oils decreased levels of low density lipoprotein cholesterol or "bad" cholesterol. That was not the case for coconut oil.[2]

In a study published in 2011 in the *Asian Pacific Journal of Clinical Nutrition*, researchers from the Philippines and Chapel Hill, North Carolina, examined the association between the consumption of coconut oil and the lipid profiles of a cohort of 1,839 Filipino women. All of the women were participants in the Cebu Longitudinal Health and Nutrition Survey, a community based survey conducted in Cebu City, Philippines. The researchers found that the intake of coconut oil had a "positive association" with total cholesterol and high-density lipoprotein (HDL). This was particularly true for premenopausal women. So, the women with high intakes of coconut oil had reductions in total cholesterol and increases in HDL. Though the increases in HDL were relatively "modest," the researchers noted that they are "likely to have clinical relevance at the population level." Moreover, "high coconut oil consumption predicted a statistically significant in-crease in HDL-c."[3]

In a study published in 2010 in the *Journal of Atherosclerosis and Thrombosis*, researchers from India examined the effect of coconut oil and sunflower oil on the risk factors for coronary artery disease in New Zealand white rabbits. The re-searchers began by dividing their rabbits into four groups. The six rabbits in the first group ate only rabbit food; they were the controls. The five rabbits in the second group were given cholesterol supplement to induce elevated cholesterol levels. The diets of the five rabbits in the third group and the five rabbits in the fourth group were supplemented with either 30% coconut oil or 30% sunflower oil. After six months, the researchers found that the coconut oil did not raise the cholesterol levels of the white rabbits.[4]

A year earlier, three of these same researchers joined with another researcher to conduct another coconut oil study, which was published in the *Indian Journal of Clinical Biochemistry.* The researchers began by placing men, between the ages of 35 and 65, into one of four groups. The first group consisted of 35 men who consumed coconut oil; the second group consisted of 35 men who consumed sunflower oil; the third group had 35 men with type 2 diabetes who consumed coconut oil; and the fourth group had 35 men with type 2 diabetes who consumed sunflower oil. All of the subjects derived about 13%–20% of their daily calories from the oils. The researchers found no significant differences between the subjects consuming coconut oil and sunflower oil. "Hence, it may be concluded that the consumption of coconut oil in moderation, as a part of a routine diet, may not contribute to the risk of CAD."[5]

In a randomized, double-blind, clinical trial published in 2009 in *Lipids,* researchers from Brazil recruited 40 women between the ages of 20 and 40 who had abdominal obesity (waist circumference of greater than 35 inches). For 12 weeks, the women took either two tablespoons per day of coconut oil or two tablespoons per day of soybean oil supplementation. Their total diets included just under 1,900 calories per day, which was less than they normally ate. The women also participated in a four-day per week exercise program directed by a fitness trainer. By the end of the trial, the women on coconut oil supplementation had significant reductions in their waist circumference. The women on soybean supplementation actually experienced increases in waist size. The researchers noted that "ingestion of coconut oil did not produce undesirable alternatives in the lipid profile of women presenting abdominal obesity, although dietary supplementation with this oil did give rise to a reduction in WC [waist circumference], which is considered to confer some protection against CVDs [cardiovascular diseases]."

The researchers also learned that the women on coconut oil supplementation had increases in HDL and reductions in C-reactive protein, a marker for inflammation. It is well known that high levels of inflammation are associated with increase risk of cardiovascular disease.[6]

IS COCONUT OIL BENEFICIAL FOR CARDIOVASCULAR SUPPORT?

From the studies presented in this entry there does appear to be some good evidence that coconut oil supports cardiovascular health. Still, before beginning a regime of supplemental coconut oil, people at higher risk for cardiovascular problems may wish to have a discussion with a medical provider.

NOTES

1. Arunima, S., and T. Rajamohan. November 2012. "Virgin Coconut Oil Improves Hepatic Lipid Metabolism in Rats—Compared with Copra Oil, Olive Oil and Sunflower Oil." *Indian Journal of Experimental Biology* 50(11): 802–809.

2. Dauqan, Eqbal, Halimah Abdullah Sani, Aminah Abdullah, and Zalifah Mohd Kasim. June 2011. "Effect of Different Vegetable Oils (Red Palm Olein, Palm Olein, Corn Oil and Coconut Oil) on Lipid Profile in Rat." *Food and Nutrition Sciences* 2(4): 253–258.

3. Feranil, A. B., P. L. Duazo, C. W. Kuzawa, and L. S. Adair. 2011. "Coconut Oil Is Associated with a Beneficial Lipid Profile in Pre-Menopausal Women in the Philippines." *Asian Pacific Journal of Clinical Nutrition* 20(2): 190–195.

4. Sabitha, P., D. M. Vasudevan, and P. Kamath. February 26, 2010. "Effect of High Fat Diet without Cholesterol Supplementation on Oxidative Stress and Lipid Peroxidation in New Zealand White Rabbits." *Journal of Atherosclerosis and Thrombosis* 17(2): 213–218.

5. Sabitha, P., K. Vaidyanathan, D. M. Vasudevan, and P. Kamath. 2009. "Comparison of Lipid Profile and Antioxidant Enzymes among South Indian Men Consuming Coconut Oil and Sunflower Oil." *Indian Journal of Clinical Biochemistry* 24(1): 76–81.

6. Assunção, M. L., H. S. Ferreira, A. F. dos Santos et al. July 2009. "Effects of Dietary Coconut Oil on the Biochemical and Anthropometric Profiles of Women Presenting Abdominal Obesity." *Lipids* 44(7): 593–601.

REFERENCES AND RESOURCES
Magazines, Journals, and Newspapers

Amarasiri, W. A. and A. S. Dissanayake. June 2006. "Coconut Fats." *The Ceylon Medical Journal* 51(2): 47–51.

Arunima, S. and T. Rajamohan. November 2012. "Virgin Coconut Oil Improves Hepatic Lipid Metabolism in Rats—Compared with Copra Oil, Olive Oil and Sunflower Oil." *Indian Journal of Experimental Biology* 50(11): 802–809.

Assunção, M. L., H. S. Ferreira, A. F. dos Santos et al. July 2009. "Effects of Dietary Coconut Oil on the Biochemical and Anthropometric Profiles of Women Presenting Abdominal Obesity." *Lipids* 44(7): 593–601.

Dauqan, Eqbal, Halimah Abdullah Sani, Aminah Abdullah, and Zalifah Mohd Kasim. June 2011. "Effect of Different Vegetable Oils (Red Palm Olein, Palm Olein, Corn Oil and Coconut Oil) on Lipid Profile in Rat." *Food and Nutrition Sciences* 4(4): 253–258.

Feranil, A. B., P. L. Duazo, C. W. Kuzawa, and L. S. Adair. 2011. "Coconut Oil Is Associated with a Beneficial Lipid Profile in Pre-Menopausal Women in the Philippines." *Asian Pacific Journal of Clinical Nutrition* 20(2): 190–195.

Sabitha, P., D. M. Vasudevan, and P. Kamath. February 26, 2010. "Effect of High Fat Diet without Cholesterol Supplementation on Oxidative Stress and Lipid Peroxidation in New Zealand White Rabbits." *Journal of Atherosclerosis and Thrombosis* 17(2): 213–218.

Sabitha, P., K. Vaidyanathan, D. M. Vasudevan, and P. Kameth. 2009. "Comparison of Lipid Profile and Antioxidant Enzymes among South Indian Men Consuming Coconut Oil and Sunflower Oil." *Indian Journal of Clinical Biochemistry* 24(1): 76–81.

Website

Coconut Oil. http://coconutoil.com.

Cod Liver Oil

Who doesn't have memories of the daily morning cod liver oil ritual? Who can forget their parents pouring spoonful after spoonful of cod liver oil? For generations of children, cod liver oil was an integral part of breakfast. Parents were convinced that cod liver oil, with its high levels of vitamins A and D, was a necessary component of the daily diet. It was believed that the consumption of cod liver oil reduced the incidence of colds and ear infections—illnesses that were so easily transmitted in day care and school environments.

But, cod liver oil, which is a rich source of omega-3 fatty acids, especially eicosapentaenoic acid and docosahexaenoic acid (DHA), is also believed to be useful for other medical problems. It is said to support cardiovascular health and reduce the pain associated with the various types of arthritis. Moreover, it is thought to be an effective treatment for psoriasis and Crohn's disease. It is even said to be useful for some psychiatric conditions including bipolar disorder, depression, and schizophrenia. Some people contend that cod liver oil should help manage the symptoms associated with autism and Asperger's syndrome.

Cod liver oil has been readily available for decades. It may be purchased in supermarkets, drugstores, discount stores, and online. Prices vary from inexpensive to expensive. Usually, it is golden in color. Try to avoid the brands that are over processed.

REDUCES THE INCIDENCE OF UPPER
RESPIRATORY INFECTIONS

In a study published in 2010 in the *Journal of the American College of Nutrition*, a New York City researcher wanted to learn if the consumption of cod liver oil by children would reduce their incidence of upper respiratory tract infections during the winter and early spring. The research was conducted at two different sites in northern Manhattan. Children between the ages of 1 and 5 received one teaspoon of Carlson's lemon-flavored cod liver oil per day and one-half tablet of Carlson's multivitamin/mineral tablet with selenium. Children between the ages of six months and one year received half that dose. The researcher found that these nutritional supplements "decreased mean visits/subject/month by 36%–58%."[1]

CARDIOVASCULAR HEALTH

In a study published in 2011 in the *European Journal of Clinical Nutrition*, researchers from Greece examined the ability of cod liver oil, extra-virgin olive oil, soy oil, and corn oil to reduce factors associated with the development of plaque within vascular walls. Two of these factors are intercellular adhesion molecule 1 (ICAM-1) and tumor necrosis factor (TNF)-alpha.

The researchers randomly placed 67 healthy volunteers with no evidence of atherosclerosis into one of four groups. Each of the participants consumed 50mL of cod liver oil, extra-virgin olive oil, soy oil, or corn oil. The researchers found that all the oils reduced ICAM-1 levels, and the cod liver oil, extra-virgin olive oil, and soy oil reduced TNF-alpha levels. They concluded that their findings "suggest that soy oil, cod liver oil, and virgin olive oil exert specific significant anti-inflammatory effects."[2]

In a meta-analysis of eight controlled trials of almost 21,000 patients published in 2009 in the *Annals of Medicine*, researchers from China found that omega-3 reduced the risk of sudden cardiac death in patients who had already experienced heart attacks. On the other hand, the researchers found that omega-3 may have adverse effects in people with angina or a type of chest pain caused by the reduced blood flow to the heart muscle.[3]

In a study published in 2012 in *Laeknabladid*, researchers from Iceland examined the association between diet and blood pressure in elderly Iceland residents. The cohort initially consisted of 99 men and 137 women between the ages of 65 and 91. For three days, the participants weighed and recorded their intake of food. One hundred and sixty of the participants correctly completed the trial. The researchers found a significant correlation between the intake of cod liver oil and lower rates of systolic blood pressure (top number—the force of blood in the arteries as the heart beats). "It is likely that cold liver oil consumption or the ingestion of supplements that contain fish oil in sufficient amounts decreases blood pressure among elderly people and could therefore have beneficial health effects."[4]

COGNITIVE IMPROVEMENTS

In a study published in 2011 in *Lipids*, researchers from Poland examined the ability of the active constituents of cod liver oil to reduce the cognitive deterioration experienced by Wistar rats undergoing chronic restraint stress. The researchers exposed rats to 2 hours of restraint stress each day for 21 days. In the follow-up testing of the rats, the researchers learned that cod liver oil "effectively restored examined cognitive functions impaired by stress." In fact, cod liver oil "significantly reduced stress-related forgetting in rats and alleviated negative effects of stress on spatial memory." The researchers concluded that "the present findings not only confirm the few to-date findings concerning the behavioral effects of DHA but also demonstrate for the first time that the use of CLO [cod liver oil] facilitates functional recovery after stress evoked cognitive deficits."[5]

ANTIULCER ACTIVITY

In a study published in 2008 in the *Indian Journal of Pharmacology*, researchers from India treated male albino Wistar rats with several different types of experimentally induced ulcers with rantidine (a medication for ulcers), low dose of cod liver oil, or high dose of cod liver oil. One group of rats was set aside to serve as the control. Though the rantidine proved to be the most effective product for healing ulcers, both doses of cod liver oil were also useful. The researchers noted that "cod liver oil may produce both gastric antisecretory and gastric cytoprotective effect, resulting in increased healing of gastric and duodenal ulcers."[6]

USEFUL FOR DEPRESSION

In a study published in 2007 in the *Journal of Affective Disorders*, researchers from Norway evaluated the association between intake of cod liver oil and the symptoms of depression in the general population. Data were obtained from a population-based cross-sectional health survey known as The Hordaland Health Study. It included close to 22,000 subjects aged 40–49 and 70–74. Among the participants in the survey, 8.9% used cod liver oil daily, and 3.6% had high levels of depressive symptoms. The researchers found that the participants who took cod liver oil were less likely to experience depression than those who did not. "The prevalence of such depressive symptoms among the subjects who used cod liver oil daily was 2.5%, as compared to 3.8% in the rest of the population."[7]

MAY HELP PEOPLE WITH RHEUMATOID ARTHRITIS

In a dual-center, double-blind, placebo-controlled randomized nine-month study published in 2008 in *Rheumatology*, researchers from the United Kingdom wanted to determine if people with rheumatoid arthritis who added cod liver oil to their diets could reduce their intake of nonsteroidal anti-inflammatory drugs (NSAID) by at least 30%. Ninety-seven patients were randomly placed in a cod liver oil supplementation group or a placebo group. After 12 weeks, the researchers asked the patients to lower or stop their NSAID intake. A total of 58 people completed the study. Nineteen of the 32 people in the cod liver group and 5 of the 26 people in the placebo group were able to reduce their NSAID requirement by 30% or more.[8]

IS COD LIVER OIL BENEFICIAL?

Obviously, cod liver oil has a number of benefits, and it is a very popular supplement. However, certain people should use cod liver oil with caution. People with cardiovascular disease, especially those on medications, should discuss cod liver oil with their health providers before beginning supplementation. Women who are pregnant or nursing should have similar discussions.

Furthermore, higher doses of cod liver oil taken over an extended period of time have the potential to be toxic. This toxicity could result in liver damage, hair loss, mental confusion, loose stools, heartburn, nausea, nosebleeds, and bone loss. To avoid these problems, it is important to take no more than the recommended doses.

NOTES

1. Linday, Linda A. December 2010. "Cod Liver Oil, Young Children, and Upper Respiratory Tract Infections." *Journal of the American College of Nutrition* 29(6): 559–562.

2. Papageorgiou, N., D. Tousoulis, T. Psaltopoulou et al. April 2011. "Divergent Anti-Inflammatory Effects of Different Oil Acute Consumption on Healthy Individuals." *European Journal of Clinical Nutrition* 65(4): 514–519.

3. Zhao, Y.T., Q. Chen, Y.X. Sun et al. 2009. "Prevention of Sudden Cardiac Death with Omega-3 Fatty Acids in Patients with Coronary Heart Disease: A Meta-Analysis of Randomized Controlled Trials." *Annals of Medicine* 41(4): 301–310.

4. Arnarson, A., O.G. Geirsdóttir, A. Ramel et al. October 2012. "Dietary Habits and Their Association with Blood Pressure among Elderly Icelandic People." *Laeknabladid* 98(10): 515–520.

5. Trofimiuk, Emil and Jan J. Braszko. 2011. "Long-Term Administration of Cod Liver Oil Ameliorates Cognitive Impairment Induced by Chronic Stress in Rats." *Lipids* 46(5): 417–423.

6. Khare, Salaj, Mohammed Asad, Sunil S. Dhamanigi, and V. Satya Prasad. 2008. "Antiulcer Activity of Cod Liver Oil in Rats." *Indian Journal of Pharmacology* 40(5): 209–214.

7. Raeder, Maria Baroy, Vidar M. Steen, Stein Emil Vollset, and Ingvar Bjelland. August 2007. "Associations between Cold Liver Oil Use and Symptoms of Depression: The Hordaland Health Study." *Journal of Affective Disorders* 101(1–3): 245–249.

8. Galarraga, B., M. Ho, H.M. Youssef et al. 2008. "Cod Liver Oil (*n-3* Fatty Acids) as an Non-Steroidal Anti-Inflammation Drug Sparing Agent in Rheumatoid Arthritis." *Rheumatology* 47(5): 665–669.

REFERENCES AND RESOURCES

Journals, Magazines, and Newspapers

Arnarson, A., O.G. Geirsdóttir, A. Ramel et al. October 2012. "Dietary Habits and Their Association with Blood Pressure among Elderly Icelandic People." *Laeknabladid* 98(10): 515–520.

Galarraga, B., M. Ho, H.M. Youssef et al. 2008. "Cod Liver Oil (*n-3* Fatty Acids) as an Non-Steroidal Anti-Inflammatory Drug Sparing Agent in Rheumatoid Arthritis." *Rheumatology* 47(5): 665–669.

Khare, Salaj, Mohammed Asad, Sunil S. Dhamanigi, and V. Satya Prasad. 2008. "Antiulcer Activity of Cold Live Oil in Rats." *Indian Journal of Pharmacology* 40(5): 209–214.

Linday, Linda A. December 2010. "Cod Liver Oil, Young Children, and Upper Respiratory Tract Infections." *Journal of the American College of Nutrition* 29(6): 559–562.

Papageorgiou, N., D. Tousoulis, A. Giolis et al. April 2011. "Divergent Anti-Inflammatory Effects of Different Oil Acute Consumption on Healthy Individuals." *European Journal of Clinical Nutrition* 65(4): 514–519.

Raeder, Maria Baroy, Vidar M. Steen, Stein Emil Vollset, and Ingvar Bjelland. 2007. "Associations between Cold Liver Oil Use and Symptoms of Depression: The Hordaland Health Study." *Journal of Affective Disorders* 101(1–3): 245–249.

Trofimiuk, Emil and Jan J. Braszko. 2011. "Long-Term Administration of Cod Liver Oil Ameliorates Cognitive Impairment Induced by Chronic Stress in Rats." *Lipids* 46(5): 417–423.

Zhao, Y. T., Q. Chen, Y. X. Sun et al. 2009. "Prevention of Sudden Cardiac Death with Omega-3 Fatty Acids in Patients with Coronary Heart Disease: A Meta-Analysis of Randomized Controlled Trials." *Annals of Medicine* 41(4): 301–310.

Website

University of Maryland Medical Center. www.umm.edu.

Evening Primrose Oil

Obtained from the seeds of the evening primrose plant, evening primrose oil is rich in omega-6 fatty acids, such as linoleic acid and gamma-linolenic acid. As a result, it is thought to be useful for medical problems characterized by inflammation, such as atopic dermatitis and rheumatoid arthritis. It has also been used for breast pain (mastalgia) as well as premenstrual and menopausal symptoms.[1]

Evening primrose oil is readily available in traditional markets as well as health stores. It is also available online. It is moderately priced. It is not uncommon for people to buy evening primrose oil in higher doses.

MAY HELP MANAGE BREAST PAIN (MASTALGIA)

In a randomized, double-blind, placebo-controlled study published in 2010 in *Alternative Medicine Review*, researchers from Minnesota examined the ability of evening primrose oil, vitamin E, and a combination of evening primrose oil and vitamin E to reduce breast pain. The initial cohort consisted of 85 women with premenstrual cyclical breast discomfort. They were assigned to one of four six-month oral treatments—evening primrose oil (3,000mg/day), vitamin E (1,200 IU/day), evening primrose and vitamin E at the same doses, or placebos. Only 41 women completed the study. Still, the results are notable. The researchers found that "daily doses of 1,200 IU vitamin E, 3,000mg EPO [evening primrose oil], or vitamin E and EPO in combination at these same dosages taken for six months may decrease the severity of cyclical mastalgia."[2]

In a study published in 2010 in *The Internet Journal of Surgery*, researchers from India noted that about 70% of women experience breast pain at some point in their lives; it is clearly a very common medical problem. To examine the effectiveness of evening primrose oil, they recruited a cohort of 89 women between the ages of 17 and 40 with mild to severe breast pain. The women were advised to eat a low-fat diet and were provided with evening primrose oil supplementation, 1000mg/day for six months. Eleven women did not complete the trial. The researchers found that evening primrose oil was more effective for the milder forms of breast pain. "Evening primrose oil can be used as first line of treatment in mild to moderate mastalgia; however, its efficacy for moderate to severe mastalgia remains doubtful."[3]

Meanwhile, in a meta-analysis published in 2007 in *The Breast*, researchers from the United Kingdom and India reviewed randomized, controlled English language trials on the use of evening primrose oil and three other products (Bromocriptine, Danazol, and Tamoxifen) for cyclical mastalgia. While Bromocriptine, Danazol, and Tamoxifen, which alter certain types of hormone levels in the blood, were all found to "offer significant relief from mastalgia," the researchers noted that evening primrose oil "is ineffective and should not be used."[4]

MAY BE USEFUL FOR ATOPIC DERMATITIS

In a study published in 2008 in the *Indian Journal of Dermatology, Venereology & Leprology*, researchers based in India investigated the use of evening primrose oil for atopic dermatitis, a chronic skin disorder characterized by scaly and itchy rashes. The initial cohort consisted of 69 patients. Of these, 29 were placed on evening primrose oil and 36 were placed on a placebo. However, only 26 people in the evening primrose group and 27 people in the placebo group completed the study. The researchers decided to analyze the results based on the first 25 people in each group who completed the five-month trial. The researchers found that at the end of the trial 96% of the people in the evening primrose group and 32% of the people in the placebo group improved. "There was significant difference in outcome of treatment between the two groups."[5]

MAY BE HELPFUL FOR PEOPLE WITH DYSLEXIA

In an open pilot study published on 2007 in the *Journal of Medicinal Food*, researchers from Sweden and the United Kingdom wanted to determine if evening primrose oil could help children with dyslexia, a developmental reading disorder in which the brain has a specific information processing problem. The cohort consisted of 20 Swedish children with dyslexia. For five months, they all took eight capsules per day of supplements containing evening primrose oil and a high-DHA fish oil. During the trial, several subjective assessments were completed by the children and their parents. The researchers found that evening primrose oil was associated with improvements in reading speed, general schoolwork, and "overall perceived benefit." They concluded that the supplement "provided positive and clear beneficial effect on variables usually impaired by dyslexia."[6]

ANTICOAGULANT PROPERTIES

In a study published in 2009 in the *Pakistan Journal of Pharmaceutical Sciences*, researchers based in Pakistan wanted to assess the anticoagulant properties of evening primrose oil. They began with a cohort of 50 healthy white rabbits, which they divided into five groups of 10 rabbits. For two months, three of the groups received either normal, moderate, or high doses of evening primrose oil. One group received warfarin, a blood thinner. And, the final group, the control, had water. Blood samples were taken after 30 days and again after 60 days.

The researchers learned that evening primrose oil demonstrated strong anticoagulant activity. In fact, the "response of [the] moderate dose was almost equivalent to the response of warfarin." They concluded that evening primrose oil "has anticoagulant properties and its anticoagulant activity is supported by its anti-inflammatory effect." Moreover, "the effects along with antiplatelet activity suggest that EPO may be of value in cardiovascular diseases."[7]

MAY REDUCE RISK OF STONE FORMATION
IN THE KIDNEYS, BLADDER, OR URETHRA

In a study published in 2009 in *The Journal of Urology*, researchers from South Africa and Italy wanted to learn if evening primrose oil could reduce the formation of stones, which primarily contain calcium oxalate, in the kidneys, bladder, and urethra, an exceedingly painful condition known as urolithiasis. For 20 days, eight black and eight white healthy male subjects ate their normal diets; they also took a daily supplement of 1,000mg evening primrose oil. The researchers found that the consumption of evening primrose oil increased the amount of citric acid in the urine. They also found that evening primrose oil may lower the risk of the development of these stones. In fact, they commented that their results "provide compelling evidence to support the notion that EPO supplementation has the potential to be a conservative therapeutic modality CaOx urolithiasis."[8]

MAY HELP WITH SYMPTOMS OF ATTENTION-DEFICIT
HYPERACTIVITY DISORDER

In a study published in 2012 in the *Journal of Child Neurology*, researchers from Sri Lanka investigated whether evening primrose oil could help children who suffer from the symptoms of attention-deficit hyperactivity disorder (ADHD), such as the inability to concentrate. The cohort consisted of 94 children between the ages of 6 and 12 who had been diagnosed with ADHD. All of the children had been treated with standard behavior therapy and had taken the ADHD medication Ritalin for at least six months. Yet, the parents observed no improvement "in behavior and academic learning." The children were randomly assigned to take evening primrose oil and fish oil supplements or a placebo. Though no significant improvements were noted after three months, after six months, parents and teachers reported significant improvements in inattention, impulsiveness, and cooperation. Evening primrose oil did not appear to alter the children's level of distractibility.[9]

MAY BE USEFUL FOR CONTACT LENS-ASSOCIATED DRY EYE

In a study published in 2008 in *Contact Lens & Anterior Eye*, researchers from the United Kingdom noted that it is not uncommon for people wearing contact lenses to feel a "sensation of dryness." Apparently, it is a frequent cause of people reducing the amount of time that they wear their contact lenses or making the decision to stop using them altogether. So, the researchers decided to evaluate the effect that the oral treatment of evening primrose oil may have on eye dryness. The initial cohort consisted of 76 females who all wore contact lenses and experienced dryness in their eyes. They were randomly placed in an evening primrose group or a placebo group. For six months, everyone took six capsules per day. Fifty-two women completed the trial. The researchers learned that evening primrose oil improved the contact lens–related eye dryness and increased tear production. "In conclusion, this study provides evidence for a beneficial effect of particular orally administered omega-6 fatty acids in alleviating symptoms and improving overall lens comfort in patients suffering from contact lens associated dry eye."[10]

IS EVENING PRIMROSE OIL BENEFICIAL?

For some people with certain conditions, evening primrose oil may well be beneficial. For example, in the previously noted instance of contact lens–related eye dryness, it is certainly worth experimenting with evening primrose oil supplementation. It may well be a good idea to discuss evening primrose oil with your medical provider.

NOTES

1. Bayles, Bryan and Richard Usatine. December 15, 2009. "Evening Primrose Oil." *American Family Physician* 80(12): 1405–1408.

Bhatt, K., M. Lodhia, and V. Thaker. March–April 2009. "Antibacterial Activity of Essential Oils from Palmarosa, Evening Primrose, Lavender and Tuberose." *Indian Journal of Pharmaceutical Sciences* 71(2): 134–136.

2. Pruthi, Sandhya, Dietlind L. Wahner-Roedler, Carolyn J. Torkelson et al. 2010. "Vitamin E and Evening Primrose Oil for Management of Cyclical Mastalgia: A Randomized Pilot Study." *Alternative Medicine Review* 15(1): 59–67.

3. Thakur, Natasha, Babar Rashid Zargar Zargar, Nadeem U.I. Nazeer et al. July 20, 2010. "Mastalgia—Use of Evening Primrose Oil in Treatment of Mastalgia." *The Internet Journal of Surgery* 24(2): NA.

    4. Srivastava, A., R. E. Mansel, N. Arvind et al. 2007. "Evidence-Based Management of Mastalgia: A Meta-Analysis of Randomised Trials." *The Breast* 16(5): 503–512.

    5. Senapati, Swapan, Sabyasachi Banerjee, and Dwijendra Nath Gangopadhyay. September–October 2008. "Evening Primrose Oil Is Effective in Atopic Dermatitis: A Randomized Placebo-Controlled Trial." *Indian Journal of Dermatology, Venereology & Leprology* 74(5): 447–452.

    6. Lindmark, Lars and Peter Clough. December 2007. "A 5-Month Open Study with Long-Chain Polyunsaturated Fatty Acids in Dyslexia." *Journal of Medicinal Food* 10(4): 662–666.

    7. Riaz, A., R. A. Khan, and S. P. Ahmed. October 2009. "Assessment of Antico-agulant Effect of Evening Primrose Oil." *Pakistan Journal of Pharmaceutical Sciences* 22(4): 355–359.

    8. Rodgers, Allen, Sonja Lewandowski, Shameez Allie-Hamdulay et al. December 2009. "Evening Primrose Oil Supplementation Increases Citraturia and Decreases Other Urinary Risk Factors for Calcium Oxalate Urolithiasis." *The Journal of Urology* 182(6): 2957–2963.

    9. Perera, Hemamali, Kamal Chandima Jeewandara, Sudarshi Seneviratne, and Chandima Guruge. June 2012. "Combined ω3 and ω6 Supplementation in Children with Attention-Deficit Hyperactivity Disorder (ADHD) Refractory to Methylphenidate Treatment: A Double-Blind, Placebo-Controlled Study." *Journal of Child Neurology* 27(6): 747–753.

    10. Kokke, K. H., J. A. Morris, and J. G. Lawrenson. June 2008. "Oral Omega-6 Essential Fatty Acid Treatment in Contact Lens Associated Dry Eye." *Contact Lens & Anterior Eye* 31(3): 141–146.

REFERENCES AND RESOURCES

Magazines, Journals, and Newspapers

Bayles, Bryan and Richard Usatine. December 15, 2009. "Evening Primrose Oil." *American Family Physician* 80(12): 1405–1408.

Bhatt, K., M. Lodhia, and V. Thaker. March–April 2009. "Antibacterial Activity of Essential Oils from Palmarosa, Evening Primrose, Lavender and Tuberose." *Indian Journal of Pharmaceutical Sciences* 71(2): 134–136.

Kokke, K. H., J. A. Morris, and J. G. Lawrenson. June 2008. "Oral Omega-6 Essential Fatty Acid Treatment in Contact Lens Associated Dry Eye." *Contact Lens & Anterior Eye* 31(3): 141–146.

Lindmark, Lars and Peter Clough. December 2007. "A 5-Month Open Study with Long-Chain Polyunsaturated Fatty Acids in Dyslexia." *Journal of Medicinal Food* 10(4): 662–666.

Perera, Hemamali, Kamal Chandima Jeewandara, Sudarshi Seneviratne, and Chandima Guruge. June 2012. "Combined ω3 and ω6 Supplementation in Children with Attention-Deficit Hyperactivity Disorder (ADHD) refractory to Methylphenidate Treatment: A Double-Blind, Placebo-Controlled Study." *Journal of Child Neurology* 27(6): 747–753.

Pruthi, Sandhya, Dietlind L. Wahner-Roedler, Carolyn J. Torkelson et al. 2010. "Vitamin E and Evening Primrose Oil for Management of Cyclical Mastalgia: A Randomized Pilot Study." *Alternative Medicine Review* 15(1): 59–67.

Riaz, A., R. A. Khan, and S. P. Ahmed. October 2009. "Assessment of Anticoagulant Effects of Evening Primrose Oil." *Pakistan Journal of Pharmaceutical Sciences* 22(4): 355–359.

Rodgers, Allen, Sonja Lewandowski, Shameez Allie-Hamdulay et al. December 2009. "Evening Primrose Supplementation Increases Citraturia and Decreases Other Urinary Risk Factors for Calcium Oxalate Urolithiasis." *The Journal of Urology* 182(6): 2957–2963.

Senapati, Swapan, Sabyasachi Banerjee, Dwijendra Nath Gangopadhyay. September–October 2008. "Evening Primrose Oil Is Effective in Atopic Dermatitis: A Randomized Placebo-Controlled Trial." *Indian Journal of Dermatology, Venereology & Leprology* 74(5): 447–452.

Srivastava, A., R. E. Mansel, N. Arvind et al. 2007. "Evidence-Based Management of Mastalgia: A Meta-Analysis of Randomised Trials." *The Breast* 16(5): 503–512.

Thakur, Natasha, Babar Rashid Zargar Zargar, Nadeem U. I. Nazeer et al. July 20, 2010. "Mastalgia—Use of Evening Primrose Oil in Treatment of Mastalgia." *The Internet Journal of Surgery* 24(2): NA.

Websites

Memorial Sloan-Kettering Cancer Center. www.mskcc.org.
University of Maryland Medical Center. www.umm.edu.

Fish Oil: Overview

It is relatively easy to include fish oil in the diet. Several fish have abundant supplies of these oils known as omega-3 fatty acids. These include salmon, black cod, mackerel, bluefish, anchovies, sardines, herring, and trout. But, fish oil is also available as a supplement. In fact, it is one of the most widely used supplements. Fish oil supplements are generally made from salmon, mackerel, herring, tuna, halibut, whale or seal blubber, or cod liver.

Why is fish oil so popular? Fish oil is believed to be useful for a wide variety of medical problems. These include cardiovascular concerns such as high blood pressure and elevated levels of triglycerides, problems with memory, glaucoma, age-related macular degeneration, breast pain, diabetes, asthma, dyslexia, obesity, kidney disease, osteoporosis, and skin problems such as psoriasis.[1]

Fish oil is sold in a wide variety of stores, and it is also readily available online. Prices for fish oil range from relatively inexpensive to expensive. If you decide to begin a fish oil regime, you may wish to discuss your plans with your medical provider.

MAY BE USEFUL IN THE PREVENTION OF
GUM (PERIODONTAL) DISEASE

In a study published in 2010 in the *Journal of the American Dietetic Association*, researchers based in Boston, Massachusetts (United States), wondered if the n-3 fatty acids, such as those found in fish oil, would be useful in the prevention and/or treatment of gum disease. The cohort consisted of 9,182 adults aged 20 years and older who participated in the National Health and Nutrition Examination Survey between 1999 and 2004. The researchers were able to detect gum disease through dental examinations, and the intake of n-3 fatty acids was determined during interviews with the subjects. Although the researchers found that 8.2% of the subjects had gum disease, the incidence was 20% lower among those with the top one-third intake of docosahexaenoic acid (DHA), found in fish oil. For those with the top one-third intake of eicosapentaenoic acid (EPA), found in fish oil, the risk was about 15% lower. Thus, the consumption of DHA and EPA is inversely associated with the risk of gum disease; those who consume the most DHA and EPA are the least likely to have gum disease.[2]

In another study published in 2010 in the *Journal of Periodontology*, researchers from Egypt and Boston examined the use of omega-3 polyunsaturated fatty acids, such as fish oil, and aspirin for advanced chronic gum disease. The cohort consisted of 80 healthy subjects between the ages of 32 and 66. Everyone received in-office treatment for their gum disease, a procedure known as scaling and root planing. During this procedure, the dentist cleaned between gums and teeth and down into the roots. After the procedure, the control group of 40 received a placebo and the treatment group of 40 was provided with fish oil supplementation and aspirin. The trial continued for 24 weeks. The researchers found that when compared to control group, the subjects who took fish oil supplementation and aspirin had significant reduction in the gum disease. They commented that this treatment "may provide a sustainable, low-cost intervention to augment periodontal therapy."[3]

MAY OR MAY NOT HELP INFLAMMATORY ACNE

In a study published in 2012 in *Lipids in Health and Disease*, researchers from Pomona, California (United States), wanted to learn if fish oil could help 13 healthy males between the ages of 18 and 40 with inflammatory acne. For 12 weeks, everyone took fish oil capsules. The researchers found that eight men had improvements in their acne and four found that their acne worsened. One man's acne remained about the same. Among those who improved, seven had moderate to severe acne at baseline. Three of the four who experienced acne that worsened had mild acne. While the researchers acknowledged that their sample was small and without a placebo group, they still noted that fish oil may be useful for people with acne, "especially for individuals with moderate and severe acne."[4]

MAY HELP PREVENT PRESSURE ULCERS

In a study published in 2012 in the *American Journal of Critical Care*, researchers from Israel assessed the ability of a feeding formula enriched with fish oil to assist the healing of pressure ulcers in critically ill patients. According to the researchers, critically ill patients, who are more likely to be inactive, are at increased risk for pressure ulcers, which are painful and might easily become infected. So, it is important to try and prevent them whenever possible. Would fish oil help? The cohort consisted of 40 critically ill patients. Half were given the usual formula and the other half had a formula enriched with fish oil and micronutrients. The researchers found that the patients on the enriched formula "had significantly less progression of existing pressure ulcers." They also had lower levels of serum C-reactive protein, a measure of inflammation.[5]

MAY HELP PEOPLE LOSE WEIGHT

In a study published in 2007 in the *International Journal of Obesity*, researchers from Iceland, Spain, Portugal, and the Netherlands wanted to determine if including seafood and fish oil in an energy-restricted diet would help overweight men and women lose weight. The cohort consisted of 324 men and women between the ages of 20 and 40. For eight weeks, the subjects were randomly placed on one of four diets. The subjects in one group ate no seafood and took placebo capsules, the subjects in a second group ate lean fish three times each week, the subjects in a third group ate fatty fish three times per week, and the subjects in the fourth group took six capsules of fish oil each day. By the end of the trial there were 278 subjects. All of the members of the three treatment groups lost more weight than the subjects who ate no seafood and took placebos. And, the men lost more weight than the women. The researchers noted that including some form of seafood in a "nutritionally balanced energy-deficient diet may boost weight loss."[6]

MAY HELP THE BODY ADD LEAN MASS
AND DECREASE FAT MASS

In a study published in 2010 in the *Journal of the International Society of Sports Nutrition*, researchers from Gettysburg, Pennsylvania (United States), investigated the use of fish oil to add lean mass and reduce fat mass from the body. The initial cohort consisted of 47 healthy adults between the ages of 18 and 55. They were randomly assigned to take either fish oil or safflower oil supplementation for six weeks. The subjects on the fish oil experienced significant increases in lean mass and significant reductions of fat mass. However, the researchers acknowledged that they do not know if similar results would be obtained over a longer period of supplementation.[7]

MAY ELEVATE POSTEXERCISE IMMUNITY

In a randomized, double-blind placebo-controlled study published in 2012 in *Brain, Behavior, and Immunity*, researchers from the United Kingdom noted that exercise may alter the immune functioning of the body. They wondered if fish oil could prevent such changes. Their cohort consisted of 16 male volunteers who were recreationally active. For six weeks, eight males took fish oil supplementation and eight took a placebo. Then, they all cycled for one hour. Blood tests were taken both before and after the cycling was completed. The researchers found that the men who had taken fish oil had an increase in natural killer cell activity. And, they suggested that this might improve their resistance to infection during the postexercise period.[8]

MAY HELP MEN WITH FERTILITY PROBLEMS

In a study published in 2010 in *Clinical Nutrition*, researchers from Iran wanted to determine if fish oil would be useful for men struggling with infertility. The cohort consisted of 82 men who were infertile and 78 men who were fertile. The researchers found that the fertile men had higher blood and sperm levels of omega-3 fatty acids, and the infertile men had significantly higher blood ratios of omega-6 to omega-3 fatty acids. The researchers wrote that their findings suggest "the beneficial effect of the higher concentrations of omega-3 FAs [fatty acids] in blood plasma and spermatozoa on semen parameters, as well as the anti-oxidant status of seminal plasma."[9]

IS FISH OIL BENEFICIAL?

Fish oil does appear to be at least somewhat effective for a wide variety of medical concerns. Still, before beginning a fish oil regime, it is a good idea to have a discussion with your medical provider.

NOTES

1. WebMD. www.webmd.com.

2. Naqvi, A.Z., C. Buettner, R.S. Phillips et al. November 2011. "n-3 Fatty Acids and Periodontitis." *Journal of the American Dietetic Association* 110(11): 1669–1675.

3. El-Sharkawy, H., N. Aboelsaad, M. Eliwa et al. November 2010. "Adjunctive Treatment of Chronic Periodontitis with Daily Dietary Supplementation with Omega-3 Fatty Acids and Low-Dose Aspirin." *Journal of Periodontology* 81(11): 1635–1643.

4. Khayef, Golandam, Julia Young, Bonny Burns-Whitmore, and Thomas Spalding. December 2012. "Effects of Fish Oil Supplementation on Inflammatory Acne." *Lipids in Health and Disease* 11(1): 165+.

5. Theilla, M., B. Schwartz, J. Cohen et al. 2012. "Impact of a Nutritional Formula Enriched in Fish Oil and Micronutrients on Pressure Ulcers in Critical Care Patients." *American Journal of Critical Care* 21(4): e102–e109.

6. Thorsdottir, I., H. Tomasson, I. Gunnarsdottir et al. 2007. "Randomized Trial of Weight-Loss-Diets for Young Adults Varying in Fish and Fish Oil Content." *International Journal of Obesity* 31(10): 1560–1566.

7. Noreen, E. E., M. J. Sass, M. L. Crowe et al. October 8, 2010. "Effects of Supplemental Fish Oil on Resting Metabolic Rate, Body Composition, and Salivary Cortisol in Healthy Adults." *Journal of the International Society of Sports Nutrition* 7: 31+.

8. Gray, Patrick, Brendan Gabriel, Frank Thies, and Stuart R. Gray. 2012. "Fish Oil Supplementation Augments Post-exercise Immune Function in Young Males." *Brain, Behavior, and Immunity* 26(8): 1265–1272.

9. Safarinejad, Mohammad Reza, Seyyed Yousof Hosseini, Farid Dadkhah, and Majid Ali Asgari. 2010. "Relationship of Omega-3 and Omega-6 Fatty Acids with Semen Characteristics, and Anti-Oxidant Status of Seminal Plasma: A Comparison between Fertile and Infertile Men." *Clinical Nutrition* 29(1): 100–105.

REFERENCES AND RESOURCES

Magazines, Journals, and Newspapers

El-Sharkawy, H., N. Aboelsaad, M. Eliwa et al. November 2010. "Adjunctive Treatment of Chronic Periodontitis with Daily Dietary Supplementation with Omega-3 Fatty Acids and Low-Dose Aspirin." *Journal of Periodontology* 81(11): 1635–1643.

Gray, Patrick, Brendan Gabriel, Frank Thies, and Stuart R. Gray. 2012. "Fish Oil Supplementation Augments Post-exercise Immune Function in Young Males." *Brain, Behavior, and Immunity* 26(8): 1265–1272.

Khayef, Golandam, Julia Young, Bonny Burns-Whitmore, and Thomas Spalding. December 2012. "Effects of Fish Oil Supplementation on Inflammatory Acne." *Lipids in Health and Disease* 11(1): 165+.

Naqvi, A. Z., C. Buettner, R. S. Phillips et al. November 210. "n-3 Fatty Acids and Periodontitis in US Adults." *Journal of the American Dietetic Association* 110(11): 1669–1675.

Noreen, E. E., M. J. Sass, M. L. Crowe et al. October 8, 2010. "Effects of Supplemental Fish Oil on Resting Metabolic Rate, Body Composition, and Salivary Control in Healthy Adults." *Journal of the International Society of Sports Nutrition* 7: 31+.

Safarinejad, Mohammad Reza, Seyyed Yousof Hosseinu, Farid Dadkhah, Majid Ali Asgari. 2010. "Relationship of Omega-3 and Omega-6 Fatty Acids with Semen Characteristics, and Anti-Oxidant Status of Seminal Plasma: A Comparison between Fertile and Infertile Men." *Clinical Nutrition* 29(1): 100–105.

Theilla, M., B. Schwartz, J. Cohen et al. July 2012. "Impact of a Nutritional Formula Enriched in Fish Oil and Micronutrients on Pressure Ulcers in Critical Care Patients." *American Journal of Critical Care* 21(4): e102–e109.

Thorsdottir, I., H. Tomasson, I. Gunnarsdottir et al. 2007. "Randomized Trial of Weight-Loss-Diets for Young Adults Varying in Fish and Fish Oil Content." *International Journal of Obesity* 31(10): 1560–1566.

Websites

Mayo Clinic. www.mayoclinic.com.
WebMD. www.webmd.com.

Fish Oil and Aging-Related Issues

As people enter their senior years, they often wonder if they should include fish oil supplementation in their diets. Would adding fish oil help defer or completely prevent so many of the problems that may be associated with aging, such as memory loss and muscle deterioration? Different medical providers offer varying opinions on the topic. But, there is interesting research.

IMPROVEMENTS IN MEMORY

In a crossover, placebo-controlled study published in 2012 in *Nutrition Journal*, researchers from Sweden recruited 40 healthy men and women between the ages of 51 and 72. For five weeks, the subjects took either daily fish oil supplementation or a placebo. After a five-week washout period, they took the other product for an additional five weeks. The researchers found that the subjects taking fish oil did better on tests that measured working memory and selective attention. And, they concluded that fish oil improved cognitive functioning in their healthy middle-aged and elderly subjects.[1]

In a study published in 2004 in *The American Journal of Clinical Nutrition*, researchers from the United Kingdom gathered data from 350 men and women who had participated in a national IQ survey in 1947 when they were 11. The researchers tested them again in 2000 and 2001 when they were 64. The researchers found that the cognitive functioning of the subjects who took supplements, especially fish oil, were higher than those who did not take supplements. They concluded that people who take fish oil have the potential to "improve retention of cognitive function in old age."[2]

In a study published in 2012 in *PLoS ONE*, researchers based in Pittsburgh, Pennsylvania (United States), began with a cohort of 13 healthy adults with an average age of 22. For six months, all the participants took 2g of fish oil per day. Eleven participants completed the entire study. When given an important challenge to working memory, they improved by 23%. The researchers hypothesized that fish oil may increase cognition by reducing inflammation or boosting the signals that pass among brain cells or maximize the availability of the neurotransmitter dopamine to other parts of the brain.[3]

In a meta-analysis that was published in 2012 in *The Journal of Clinical Psychiatry*, researchers from Taiwan wanted to learn if people with dementia or pre-dementia had altered levels of fatty acids found in fish oil, such as docosahexaenoic acid (DHA) and eicosapentaenoic (EPA). Their study search yielded 10 articles, which included 2,280 subjects. The researchers found that the levels of DHA and EPA were lower in people with dementia, and the levels of EPA but not DHA were lower in people with pre-dementia. And, the researchers added that their work supports the important role that n-3 polyunsaturated fatty acids, such as those found in fish oil, play in the pathophysiology of dementia. In addition, "EPA might be not only a disease-state marker but also a risk factor for cognitive impairment."[4]

In a 12-month, randomized, double-blind, placebo-controlled study published in 2012 in *Psychopharmacology*, researchers from Malaysia wanted to determine if fish oil would be useful for people already dealing with mild cognitive impairment (MCI). Thirty-six elderly subjects with MCI were assigned to receive either concentrated fish oil or a placebo. The 18 members of the fish oil groups showed significant improvement in short-term and working memory, immediate verbal memory, and delayed recall capability. The researchers noted that the change in memory "was significantly better in the fish oil group."[5]

And, in a study published in 2013 in the *British Journal of Nutrition*, researchers from Saint Louis, Missouri (United States), and Toronto, Canada, supplemented the diet of some of their middle-aged to elderly cats (5.5–8.7 years) with fish oil, B vitamins, antioxidants, and arginine. Meanwhile, the control cats ate their regular diet; they did not receive any of the supplementation. Then, the researchers conducted cognitive tests on the cats. They found that the cats on the supplementation performed better in three of the four test protocols. As a result, the researchers concluded that the nutrient blend they selected reduced or eliminated some of the cognitive problems associated with the brains of aging cats.[6]

On the other hand, in a study published in 2010 in the *American Journal of Clinical Nutrition*, researchers from the United Kingdom and Australia tested fish oil supplementation on 867 cognitively healthy adults between the ages of 70 and 79. For 24 months, the subjects were randomly assigned to take daily fish oil or olive oil capsules. Cognition tests were administered by trained research nurses at baseline and at the end of the trial. Seven hundred and forty-eight participants completed the study. But, there were no significant differences between the groups, and neither group showed a decline in cognition function.[7]

ENHANCES STRENGTH TRAINING

In a study published in 2012 in *The American Journal of Clinical Nutrition*, researchers from Brazil examined the effect of fish oil supplementation and strength training on the neuromuscular system of older women. The researchers randomly assigned 45 women around the age of 64 to one of three groups. The women in the first group performed strength training three times each week for 90 days. The women in the second group performed the same strength training and took fish oil supplementation for 90 days. And, the women in the last group began taking fish oil supplementation 60 days before they started training, and then combined strength training with supplementation for an additional 90 days. The researchers learned that fish oil supplementation combined with strength training "improved the response of the neuromuscular system." As a result, fish oil "may be an attractive supplement for the elderly." However, the researchers added that they did not observe any difference between the subjects who took the fish oil supplementation for 90 days and those who took it for 150 days.[8]

MAY HELP MAINTAIN MUSCLE MASS

In a randomized, controlled study published in 2011 in *The American Journal of Clinical Nutrition*, researchers from Saint Louis, Missouri (United States), and the United Kingdom noted that the loss of muscle mass that so often occurs as people age is a "major public health concern." When people lose muscle mass, their quality of life may be diminished, and they are at increased risk of becoming frail. Could fish oil help them maintain muscle mass? The researchers assembled a cohort of 10 men and 6 women, aged 65 or older. For eight weeks, they took either fish oil or corn oil supplementation. Fifteen subjects completed the trial. While the researchers found that corn oil had no effect on the rate of muscle protein synthesis, the fish oil did support this process. As a result, they concluded that fish oil may help older adults retain their muscle mass.[9]

IS FISH OIL BENEFICIAL FOR SOME AGING-RELATED PROBLEMS?

Fish oil may well be useful for some aging-related problems. However, before beginning a fish oil regime, it is a good idea to discuss the issue with a medical provider.

NOTES

1. Nilsson, A., K. Radeborg, I. Salo, and I. Björck. November 22, 2012. "Effects of Supplementation with *n*-3 Polyunsaturated Fatty Acids on Cognitive Performance and Cardiometabolic Risk Markers in Healthy 51 to 72 Years Old Subjects: A Randomized Controlled Cross-Over Study." *Nutrition Journal* 11(1): 99+.

2. Whalley, Lawrence J., Helen C. Fox, Klaus W. Wahle et al. December 2004. "Cognitive Aging, Childhood Intelligence, and the Use of Food Supplements: Possible Involvement of *n*-3 Fatty Acids." *The American Journal of Clinical Nutrition* 80(6): 1650–1657.

3. Narendran, R., W. G. Frankle, N. S. Mason et al. October 3. 2012. "Improved Working Memory but No Effect on Striatal Vesicular Monoamine Transporter Type 2 after Omega-3 Polyunsaturated Fatty Acid Supplementation." *PLoS ONE* 7(10): e46832.

4. Lin, P. Y., C. C. Chiu, S. Y. Huang, and K. P. Su. September 2012. "A Meta-Analytic Review of Polyunsaturated Fatty Acid Compositions in Dementia." *The Journal of Clinical Psychiatry* 73(9): 1245–1254.

5. Lee, L. K., S. Shahar, A. V. Chin, and N. A. Yusoff. February 2013. "Docosahexaenoic Acid-Concentrated Fish Oil Supplementation in Subjects with Mild Cognitive Impairment (MCI): A 12-Month Randomised, Double-Blind, Placebo-Controlled Trial." *Psychopharmacology* 225(3): 605–612.

6. Pan, Y., J. A. Araujo, J. Burrows et al. July 2013. "Cognitive Enhancement in Middle-Aged and Old Cats with Dietary Supplementation with a Nutrient Blend Containing Fish Oil, Vitamins, Antioxidants and Arginine." *British Journal of Nutrition* 110(1): 40–49.

7. Dangour, Alan D., Elizabeth Allen, Diana Elbourne et al. June 2010. "Effect of 2-Y *n*-3 Long-Chain Polyunsaturated Fatty Acid Supplementation on Cognitive Function in

Older People: A Randomized, Double-Blind, Controlled Trial." *The American Journal of Clinical Nutrition* 91(6): 1725–1732.

8. Rodacki, Cintia L.N., André L.F. Rodacki, Gleber Pereira et al. February 2012. "Fish-Oil Supplementation Enhances the Effects of Strength Training in Elderly Women." *The American Journal of Clinical Nutrition* 95(2): 428–436.

9. Smith, Gordon I., Philip Atherton, Dominic N. Reeds et al. February 2011. "Dietary Omega-3 Fatty Acid Supplementation Increases the Rate of Muscle Protein Synthesis in Older Adults: A Randomized Controlled Trial." *The American Journal of Clinical Nutrition* 93(2): 402–412.

REFERENCES AND RESOURCES

Magazines, Journals, and Newspapers

Dangour, Alan D., Elizabeth Allen, Diana Elbourne et al. June 2010. "Effect of 2-Y n-3 Fatty Acid Supplementation on Cognitive Function in Older People: A Randomized, Double-Blind, Controlled Trial." *The American Journal of Clinical Nutrition* 91(6): 1725–1732.

Lee, L.K., S. Shahar, A.V. Chin, and N.A. Yusoff. February 2013. "Docosahexaenoic Acid-Concentrated Fish Oil Supplementation in Subjects with Mild Cognitive Impairment (MCI): A 12-Month Randomised, Double-Blind, Placebo-Controlled Trial." *Psychopharmacology* 225(3): 605–612.

Lin, P.Y., C.C. Chiu, S.Y. Huang, and K.P. Su. September 2012. "A Meta-Analytic Review of Polyunsaturated Fatty Avid Compositions in Dementia." *The Journal of Clinical Psychiatry* 73(9): 1245–1254.

Narendran, R., W.G. Frankle, N.S. Mason et al. October 3, 2012. "Improved Working Memory but No Effect on Striatal Vesicular Monoamine Transporter Type 2 after Omega-3 Polyunsaturated Fatty Acid Supplementation." *PLoS ONE* 7(10): e46832.

Nilsson, A., K. Radeborg, I. Salo, and I. Björck. November 22, 2012. "Effects of Supplementation with n-3 Polyunsaturated Fatty Acids on Cognitive Performance and Cardiometabolic Risk Markers in Healthy 51 to 72 Years Old Subjects: A Randomized Controlled Cross-Over Study." *Nutrition Journal* 11(1): 99+.

Pan, Y., J.A. Araujo, J. Burrows et al. July 2013. "Cognitive Enhancement in Middle-Aged and Old Cats with Dietary Supplementation with a Nutrient Blend Containing Fish Oil, B Vitamins, Antioxidants and Arginine." *British Journal of Nutrition* 110(1): 40–49.

Rodacki, Cintia L.N., André L.F. Rodacki, Gleber Pereira et al. February 2012. "Fish-Oil Supplementation Enhances the Effects of Strength Training in Elderly Women." *The American Journal of Clinical Nutrition* 95(2): 428–436.

Smith, Gordon I., Philip Atherton, Dominic N. Reeds et al. February 2011. "Dietary Omega-3 Fatty Acid Supplementation Increases the Rate of Muscle Protein Synthesis in Older Adults: A Randomized Controlled Trial." *The American Journal of Clinical Nutrition* 93(2): 402–412.

Whalley, Lawrence, Helen C. Fox, Klaus W. Wahle et al. December 2004. "Cognitive Aging, Childhood Intelligence, and the Use of Food Supplements: Possible Involvement of n-3 Fatty Acids." *The American Journal of Clinical Nutrition* 80(6): 1650–1657.

Website

Mayo Clinic. www.mayoclinic.com.

Fish Oil and Arthritis

Fish oil is frequently said to be useful for painful inflammatory conditions, such as arthritis, especially rheumatoid arthritis. Rheumatoid arthritis is a chronic inflammatory autoimmune disease that is characterized by swelling, pain, morning stiffness, functional impairment, muscle wasting, and osteoporosis. There are a number of medications that are useful for the symptoms of rheumatoid arthritis. But, they are well known to have some problematic side effects, such as gastrointestinal bleeding and distress. So, it is not uncommon for people to try alternative or complementary treatments, such as fish oil. But, is fish oil an effective treatment?

MAY WELL BE USEFUL FOR THE PAIN
ASSOCIATED WITH ARTHRITIS

In a study published in 2009 in the *Bangladesh Medical Research Council Bulletin,* researchers from Bangladesh wanted to learn if fish oil supplementation would improve the efficacy of indomethacin, a medication used to address the pain, tenderness, swelling, and stiffness associated with rheumatoid arthritis. The cohort consisted of 100 patients with rheumatoid arthritis. They were randomly assigned to receive 12 weeks of indomethacin or 12 weeks of indomethacin and fish oil supplementation. During the weeks of treatment, 19 participants left the study. The researchers found that both indomethacin alone and in combination with fish oil were well tolerated by the subjects. Moreover, the subjects' symptoms improved with monotherapy and with combined therapy. However, "the response was significantly better in [the] combination group." The researchers commented that "the combination group demonstrated a better improvement than indomethacin alone."[1]

In a randomized, double-blind, placebo-controlled study published in 2010 in the *Journal of Parenteral and Enteral Nutrition,* researchers from Austria wanted to determine if infusions of fish oil would be useful for 24 people dealing with moderate to severe rheumatoid arthritis. For 14 consecutive days, the subjects stayed in the hospital and received infusions of fish oil or a placebo. After their discharge, they took fish oil or placebo capsules for an additional 20 weeks. Twenty people completed the infusion portion of the study; by the end of the oral capsules portion, there were only 13. Still, the researchers found a statistically significant reduction in swollen joints among the people taking intravenous fish oil. This effect continued during the weeks of oral treatment. The researches noted that the placebo treatment also "led to a considerable reduction in swollen and tender joint counts." They suggested that this may be the result of "the placebo effect as well as the intensified physician's attention and care to both groups." So, the researchers concluded that despite the large number of patients that did not complete the entire trial, it was still evident that the subjects benefited from fish oil therapy.[2]

In a meta-analysis published in 2007 in *Pain*, researchers from Toronto, Canada, examined studies on the use of fish oil for the treatment of inflammatory conditions, such as rheumatoid arthritis. Their analysis included 17 studies with a total of 823 subjects. The researchers found that after three months of fish oil supplementation, subjects have improvements in pain, "particularly with respect to patient assessed pain, duration of morning stiffness, number of painful and/ or tender joints, and NSAIM [nonsteriodal anti-inflammatory medication] consumption." They concluded that fish oil "is an attractive adjunctive treatment for joint pain."[3]

In a study published in 2012 in *Nutrition*, researchers from Japan wanted to learn if the symptoms of rheumatoid arthritis were associated with dietary habits, nutritional status, and other factors. The cohort consisted of 34 women and 3 men with rheumatoid arthritis. The researchers found correlations between the intake of monounsaturated fats, omega-3 polyunsaturated fatty acids, and fish oil and the symptoms of rheumatoid arthritis. Specifically, the intake of monounsaturated fat, omega-3 polyunsaturated fatty acids, and fish oil were associated with the suppression of the inflammatory status and lowered disease activity. They "affected disease activity and may have beneficial effects by decreasing inflammation."[4]

In a study published in 2011 in *Pharmacological Research*, researchers from San Antonio, Texas (United States), and Japan investigated the ability of fish oil to reduce the pain that mice feel when their paws are exposed to radiant heat. The researchers began by dividing 60 mice into three groups of 20. The first group of mice was fed a diet containing 4% fish oil; the second group of mice was fed a diet containing 4% concentrated fish oil; and the third group of mice was fed a diet containing 5% safflower oil. The researchers found that compared to the mice fed the safflower supplemented diet, the mice fed the concentrated fish oil supplemented diet "exhibited a modest but statistically significant reduction in sensitivity" when their paws were exposed to radiant heat. No significant effect was seen in the mice that had regular fish oil supplement. According to the researchers, "it appears that the high concentration of n-3 fatty acids, EPA and DHA in CFO [concentrated fish oil] is largely responsible for reduced pain sensitivity."[5]

In a study published in 2012 in the *American Journal of Epidemiology*, researchers from Seattle, Washington (United States), noted that it is important to identify supplements that have anti-inflammatory properties. They explained that inflammation is associated with a number of chronic medical problems. Their cohort consisted of 9,947 adults from the 1999–2004 cycles of the National Health and Nutrition Examination Survey. Measures of inflammation were determined by testing for serum high sensitivity C-reactive protein concentrations. The researchers found that fish oil was one of three supplements associated with reductions in inflammation. (The other two were glucosamine and chondroitin.)[6]

In a report published in 2012 in the *British Journal of Nutrition*, two researchers from the United Kingdom conducted a review of some of the studies on the consumption of fish oil and oily fish to treat the symptoms associated with rheumatoid arthritis, which the researchers said affects about 1% of adults. Almost all of the 23 articles included in the report found that fish oil had some clinical benefit. "Commonly reported benefits include reduced duration of morning stiffness, reduced number of tender or swollen joints, reduced joint pain, reduced time to fatigue, increased grip strength, reduced pain or disease activity assessed by physician or patient and decreased use of NSAIDs [nonsteriodal anti-inflammatory drugs]." However, the researchers acknowledged that not all of the studies found fish oil useful for rheumatoid arthritis, and they suggested that these findings may be a function of the doses of fish oil or sample sizes that are too small. There is also the possibility that some studies ended too quickly—before beneficial results could be obtained.[7]

Researchers in Sweden decided to examine the association between fish oil and rheumatoid arthritis from a different perspective. In a study published in 2009 in *Epidemiology*, these researchers wanted to determine if including fish oil and oily fish in the diet would reduce the risk of developing rheumatoid arthritis. The cohort consisted of 1,889 people with rheumatoid arthritis and 2,145 randomly recruited controls between the ages of 18 and 70. While the researchers found that the regular consumption of fish was associated with a modest decreased risk for developing rheumatoid arthritis, no such association was found between supplementation with fish oil. But, the researchers noted that "the prevalence of fish oil supplements was low and the confidence intervals broad." In addition, "misclassification of exposure due to misreporting by the participants may occur because of difficulties in recalling dietary habits." As a result, the researchers wrote, "we cannot exclude the possibility that nondifferential misclassification of fish oil supplements may explain their lack of association with risk of rheumatoid arthritis, or that regular consumption over a long period of time may be preventive."[8]

IS FISH OIL BENEFICIAL FOR ARTHRITIS, ESPECIALLY RHEUMATOID ARTHRITIS?

From the research presented, it is obvious that some people with inflammatory problems may well benefit from fish oil supplementation. Before beginning a fish oil supplementation regime, it is a good idea to discuss fish oil with a medical provider.

NOTES

1. Das Gupta, Ashoke Bijoy, A.K. Hossain, M.H. Islam et al. August 2009. "Role of Omega-3 Fatty Acid Supplementation with Indomethacin in Suppression of Disease Activity in Rheumatoid Arthritis." *Bangladesh Medical Research Council Bulletin* 35(2): 63–68.

2. Bahadori, B., E. Uitz, R. Thonhofer et al. March 2010. "ω-3 Fatty Acids Infusions as Adjuvant Therapy in Rheumatoid Arthritis." *Journal of Parental and Enteral Nutrition* 34(2): 151–155.

3. Goldberg, Robert J. and Joel Katz. May 2007. "A Meta-Analysis of the Analgesic Effects of Omega-3 Polyunsaturated Fatty Acid Supplementation for Inflammatory Joint Pain." *Pain* 129(1): 210–223.

4. Hayashi, Hitomi, Keiko Satoi, Natsuko-Sato-Mito et al. November 2012. "Nutritional Status in Relation to Adipokines and Oxidative Stress Is Associated with Disease Activity in Patients with Rheumatoid Arthritis." *Nutrition* 28(11–12): 1109–1114.

5. Veigas, J.M., P.J. Williams, G. Halade et al. May 2011. "Fish Oil Concentrate Delays Sensitivity to Thermal Nociception in Mice." *Pharmacological Research* 63(5): 377–382.

6. Kantor, Elizabeth D., Johanna W. Lampe, Thomas L. Vaughan et al. 2012. "Association between Use of Specialty Dietary Supplements and C-Reactive Protein Concentrations." *American Journal of Epidemiology* 176(11): 1002–1013.

7. Miles, Elizabeth A. and Philip C. Calder. June 2012. "Influence of Marine n-3 Polyunsaturated Fatty Acids on Immune Function and a Systematic Review of Their Effects on Clinical Outcomes in Rheumatoid Arthritis." *British Journal of Nutrition* 107(S2): S171–S184.

8. Rosell, M., A.M. Wesley, K. Rydin et al. November 2009. "Dietary Fish and Fish Oil and the Risk of Rheumatoid Arthritis." *Epidemiology* 20(6): 896–901.

REFERENCES AND RESOURCES
Magazines, Journals, and Newspapers

Bahadori, B., E. Uitz, R. Thonhofer et al. March 2010. "ω-3 Fatty Acids Infusions as Adjunct Therapy in Rheumatoid Arthritis." *Journal of Parental and Enteral Nutrition* 34(2): 151–155.

Das Gupta, Ashoke Bijoy, A.K. Hossain, M.H. Islam et al. August 2009. "Role of Omega-3 Fatty Acid Supplementation with Indomethacin in Suppression of Disease Activity in Rheumatoid Arthritis." *Bangladesh Medical Research Council Bulletin* 35(2): 63–68.

Goldberg, Robert J. and Joel Katz. May 2007. "A Meta-Analysis of the Analgesic Effects of Omega-3 Polyunsaturated Fatty Acid Supplementation for Inflammatory Joint Pain." *Pain* 129(1): 210–223.

Hayashi, Hitomi, Keiko Satoi, Natsuko Sato-Mito et al. November 2012. "Nutritional Status in Relation to Adipokines and Oxidative Stress Is Associated with Disease Activity in Patients with Rheumatoid Arthritis." *Nutrition* 28(11–12): 1109–1114.

Kantor, Elizabeth D., Johanna W. Lampe, Thomas L. Vaughan et al. 2012. "Association between Use of Specialty Dietary Supplements and C-Reactive Protein Concentrations." *Journal of Epidemiology* 176(11): 1002–1013.

Miles, Elizabeth A. and Philip C. Calder. June 2012. "Influence of Marine n-3 Polyunsaturated Fatty Acids on Immune Function and a Systematic Review of Their Effects on Clinical Outcomes in Rheumatoid Arthritis." *British Journal of Nutrition* 107(S2): S171–S184.

Rosell, M., A.M. Wesley, K. Rydin et al. November 2009. "Dietary Fish and Fish Oil and the Risk of Rheumatoid Arthritis." *Epidemiology* 20(6): 896–901.

Veigas, J.M., P.J. Williams, G. Halade et al. May 2011. "Fish Oil Concentrate Delays Sensitivity to Thermal Nociception in Mice." *Pharmacological Research* 63(5): 377–382.

Website

Arthritis Today. www.arthritistoday.org.

Fish Oil and Bone Health

There is research on the association between the consumption of fish oil and osteoporosis, a very common bone disease, especially in older women, in which bones lose too much bone or make too little bone or both. With their bones weakened, people with osteoporosis are at increased risk for fractures. Should fish oil be a useful addition to the diet of those at risk for osteoporosis and/or those already dealing with the disease?

MAY SUPPORT BONE HEALTH

In a study published in 2012 in *Bone*, researchers from France wanted to learn what would happen to the bones of aging mice that are fed diets supplemented with different fats. Using an original model of osteoporosis that mimicked the clinical features seen in humans, the researchers began by dividing their mice into four groups of 20. One group of mice was fed a standard diet. But, the other three were fed diets supplemented with sunflower oil, borage oil, or fish oil. When the mice reached 12 months, they were sacrificed, and the researchers conducted tests on their bones. The researchers found that both the diets enriched with borage oil and fish oil were able to counteract the bone loss associated with aging. On the other hand, sunflower oil "significantly exacerbated

Fish oil capsules. (Picstudio/Dreamstime.com)

bone loss." Interestingly, "sunflower associated bone loss and inflammation were counteracted by both borage and fish oil supplementation."[1]

In a study published in 2011 in *The Journal of Nutritional Biochemistry*, researchers from San Antonio, Texas (United States), examined the ability of fish oil and conjugated linoleic acid, alone or in combination, to prevent bone loss in 12-month-old mice. The researchers began by dividing 40 mice into four groups of 10. The control group of mice had their food supplemented with corn oil, a second group of mice had fish oil supplementation, a third group of mice had conjugated linoleic acid supplementation, and the final group had both fish oil and conjugated linoleic acid supplementation. The mice remained on the various diets for six months. The researchers found that the combination of fish oil and conjugated linoleic acid "helped to prevent age-associated bone loss."[2]

In a study published in 2011 in *Nutrition & Metabolism*, researchers from Iran and the University of Missouri (United States) wanted to understand the effects of long-term aerobic exercise and fish oil supplementation on postmenopausal women. The cohort consisted of 79 healthy sedentary postmenopausal women between the ages of 58 and 78. They were randomly placed in one of four groups. While 18 women were in the control group, 21 women either walked or ran and took fish oil. Another group of 20 women just walked or ran; and still another group of 20 women only took the fish oil supplementation. At the end of 24 weeks, the researchers determined that the combination of fish oil and aerobic exercise "provided numerous benefits on bone density and inflammation over exercise alone or supplementation alone." They concluded that fish oil and aerobic exercise "have a synergistic effect in attenuating inflammation and augmenting BMD [bone mineral density] in post-menopausal osteoporosis."[3]

In a study published in 2007 in *The American Journal of Clinical Nutrition*, researchers from Sweden investigated the role of fatty acids, such as fish oil, in bone accumulation and the buildup of higher amounts of bone mass in young men. The cohort consisted of 78 healthy male teens, who had a mean age of 16.7 years at baseline. Their bone mineral densities were assessed at baseline and then again at 22 and 24 years. Researchers found a positive correlation between concentrations of fatty acids and total bone mineral density and bone mineral density of the spine. "The results showed that n-3 fatty acids, especially DHA, are positively associated with bone mineral accrual and, thus, with peak BMD in young men."[4]

In a study published in 2011 in the *Journal of Bone and Mineral Metabolism*, researchers from Japan wanted to examine the effects that the intake of soy isoflavone and fish oil would have on the bone mass of ovariectomized mice. The researchers began by dividing their mice into four groups. For four weeks, the mice were fed diets supplemented with fish oil or isoflavone supplemented safflower oil or the combination of soy isoflavone and fish oil or safflower oil (control). The researchers observed that the ovariectomy significantly decreased the femoral bone mineral density. "However, this decrease was inhibited by the intake of isoflavone and/or fish oil." The researchers observed that soy isoflavone and fish oil appear to work synergistically to reduce bone resorption; they worked together better than fish oil alone.[5]

In a study published in 2012 in the *Annals of Nutrition & Metabolism*, researchers from South Korea evaluated the hypothesis that the red blood cell levels of n-3 polyunsaturated fatty acid, which is found in fish oil, and the dietary intake of fish are associated with the reduced risk of osteoporosis. The cohort consisted of 50 women with osteoporosis and 100 controls. As they expected, the researchers found that the high intake levels of n-3 polyunsaturated fatty acids and fish positively correlated with the higher bone mineral density of the femoral neck. "In particular, high erythrocyte [red blood cell] levels of EPA and DHA [found in fish oil] reduced the risk of osteoporosis in postmenopausal women after adjusting for relevant confounders."[6]

In a study published in 2008 in the *Journal of Bone and Mineral Metabolism*, researchers from Japan and Salt Lake City, Utah (United States), wanted to learn the impact that fish oil supplementation would have on the mass, density, formation, and reabsorption of bone in an aged ovariectomized rat model. Their cohort consisted of 12-month-old retired breeder Sprague-Dawley rats. For two months, they were fed a regular rat diet or a diet supplemented with fish oil. After two weeks on these diets, the rats were either sham-operated or bilaterally ovariectomized. Ten weeks later, the researchers conducted their tests on the rats. They found that the fish oil protected the rats from bone loss. "Fish oil supplementation had a positive effect on bone metabolism and might be a possible intervention to slow the loss of bone observed following menopause."[7]

And, in a study published in 2009 in the *Journal of the American Geriatrics Society*, researchers from Connecticut (United States) assessed the relationship between self-reported intake of omega-3 fatty acid, such as fish oil, and bone mineral density and lower extremity function in older adults. The initial cohort consisted of 118 men and 129 women who lived independently or in an assisted living facility in Hartford County, Connecticut. Everyone was aged 60 or older and not acutely ill. The researchers found that the men and women in the study tended to have lower intakes of omega-3 fatty acids. Still, they found an association between omega-3 fatty acid intake and higher levels of bone mineral density at the heel and hip. Subjects with lower intake of omega-3 fatty acids tended to have lower bone mineral density values; subjects with higher intakes of omega-3 fatty acids tended to have higher bone mineral density values. The researchers considered their results from this trial "promising" and suggested the need for further investigation.[8]

IS FISH OIL USEFUL FOR PEOPLE AT INCREASED RISK FOR OSTEOPOROSIS AND/OR BONE LOSS?

Fish oil may well be a useful addition to the diet of people at risk of bone loss. But, more trials in humans are clearly needed. Before beginning a fish oil regime for bone loss, it is a good idea to consult your medical provider.

NOTES

1. Wauquier, F., V. Barquissau, L. Léotoing et al. February 2012. "Borage and Fish Oil Lifelong Supplementation Decreases Inflammation and Improves Bone Health in a Murine Model of Senile Osteoporosis." *Bone* 50(2): 553–561.

2. Halade, Ganesh V., M. Rahman, Paul J. Williams, and Gabriel Fernandes. 2011. "Combination of Conjugated Linoleic Acid with Fish Oil Prevents Age-Associated Bone Marrow Adiposity in C57Bl/6J Mice." *The Journal of Nutritional Biochemistry* 22(5): 459–469.

3. Tartibian, Bakhtyar, Behzad Hajizadeh, Jill Kanaley, and Karim Sadeghi. 2011. "Long-Term Aerobic Exercise and Omega-3 Supplementation Modulate Osteoporosis Through Inflammatory Mechanisms in Post-Menopausal Women: A Randomized, Repeated Measures Study." *Nutrition & Metabolism* 8(1): 71–83.

4. Högström, M., P. Nordström, and A. Nordström. March 2007. "n-3 Fatty Acids Are Positively Associated with Peak Bone Mineral Density and Bone Accrual in Healthy Men: The NO2 Study." *The American Journal of Clinical Nutrition* 85(3): 803–807.

5. Uchida, Raina, Hiroshige Chiba, Yoshiko Ishimi et al. July 2011. "Combined Effects of Soy Isoflavone and Fish Oil on Ovariectomy-Induced Bone Loss in Mice." *Journal of Bone and Mineral Metabolism* 29(4): 404–413.

6. Moon, H. J., T. H. Kim, D. W. Byun, and Y. Park. 2012. "Positive Correlation between Erythrocyte Levels of n-3 Polyunsaturated Fatty Acids and Bone Mass in Postmenopausal Korean Women with Osteoporosis." *Annals of Nutrition & Metabolism* 60(2): 146–153.

7. Matsushita, H., J. A. Barrios, J. E. Shea, and S. C. Miller. 2008. "Dietary Fish Oil Results in a Greater Bone Mass and Bone Formation Indices in Aged Ovariectomized Rats." *Journal of Bone and Mineral Metabolism* 26(3): 241–247.

8. Rousseau, James H., Alison Kleppinger, and Anne M. Kenny. 2009. "Self-Reported Dietary Intake of Omega-3 Fatty Acids and Association with Bone and Lower Extremity Function." *Journal of the American Geriatrics Society* 57(10): 1781–1788.

REFERENCES AND RESOURCES

Magazines, Journals, and Newspapers

Halade, Ganesh V., M. Rahman, Paul J. Williams, and Gabriel Fernandes. 2011. "Combination of Conjugated Linoleic Acid with Fish Oil Prevents Age-Associated Bone Marrow Adiposity in C57Bl/6J Mice." *The Journal of Nutritional Biochemistry* 22(5):459–469.

Högström, Magnus, Peter Nordström, and Anna Nordström. March 2007. "n-3 Fatty Acids Are Positively Associated with Peak Bone Mineral Density and Bone Accrual in Healthy Men: The NO2 Study." *The American Journal of Clinical Nutrition* 85(3): 803–807.

Matsushita, H., J. A. Barrios, J. E. Shea, and S. C. Miller. 2008. "Dietary Fish Oil Results in a Greater Bone Mass and Bone Formation Indices in Aged Overiectomized Rats." *Journal of Bone and Mineral Metabolism* 26(3): 241–247.

Moon, H. J., T. H. Kim, D. W. Byun, and Y. Park. 2012. "Positive Correlation between Erythrocyte Levels of n-3 Polyunsaturated Fatty Acids and Bone Mass in Postmenopausal Korean Women with Osteoporosis." *Annals of Nutrition & Metabolism* 60(2): 146–153.

Rousseau, James H., Alison Kleppinger, and Anne M. Kenny. October 2009. "Self-Reported Dietary Intake of Omega-3 Fatty Acids and Association with Bone and Lower Extremity Function." *Journal of the American Geriatrics Society* 57(10): 1781–1788.

Tartibian, Bakhtyar, Behzad Hejizadeh, Jill Kanaley, and Karim Sadeghi. 2011. "Long-Term Aerobic Exercise and Omega-3 Supplementation Modulate Osteoporosis Through Inflammatory Mechanisms in Post-Menopausal Women: A Randomized, Repeated Measures Study." *Nutrition & Metabolism* 8(1): 71–83.

Uchida, Raina, Hiroshige Chiba, Yoshiko Ishimi et al. July 2011. "Combined Effects of Soy Isoflavone and Fish Oil on Ovariectomy-Induced Bone Loss in Mice." *Journal of Bone and Mineral Metabolism* 29(4): 404–413.

Wauquier, F., V. Barquissau, L. Léotoing et al. February 2012. "Borage and Fish Oil Life-long Supplementation Decreases Inflammation and Improves Bone Health in a Murine Model of Senile Osteoporosis." *Bone* 50(2): 553–561.

Website

National Osteoporosis Foundation. www.nof.org.

Fish Oil and Cancer

There is some interesting research on the use of fish oil to prevent various types of cancer. And, it may or may not be advised as part of a treatment protocol for cancer.

MAY BE USEFUL FOR MEN ABOUT TO UNDERGO RADICAL PROSTATECTOMY

In a trial published in 2011 in *Cancer Prevention Research,* researchers from California and North Carolina (United States) wondered if fish oil combined with a low-fat diet would be useful for men who have been diagnosed with prostate cancer and are about to undergo a radical prostatectomy. (In this surgical procedure, the prostate and surrounding tissues are removed.) The researchers randomly divided 55 men into two groups. During the four to six weeks before surgery, the men in one group ate a low-fat diet and took fish oil supplementation; the men in the other group ate their usual Western-style diet. Forty-eight men completed the study. The researchers found that the men on the low-fat/fish oil supplement diet reduced the number of rapidly dividing cells in the prostate cancer tissue. According to the researchers, that is important. It may help predict the future cancer growth. If the cancer growth is slower, it is less likely to grow beyond the prostate. Cancer that has spread beyond the gland is harder to treat.[1]

MAY INCREASE THE RISK OF PROSTATE CANCER

In a large prospective study published in 2011 in the *American Journal of Epidemiology,* researchers from Seattle, Washington (United States), analyzed data that were collected over a seven-year period on a subset of 3,400 men between the ages of 55 and 84. They found that the men with the highest concentrations of serum docosahexaenoic acid (DHA), found in fish oil, had two and a half times the risk of developing aggressive, high-grade prostate cancer than the men with the lowest DHA levels. The researchers were stunned by their results; they had expected the exact opposite to occur. They noted that their findings were "contrary to those expected from the pro- and antiinflammatory effects of these fatty acids and suggest a greater complexity of effects of these nutrients with regard to prostate cancer risk."[2]

MAY DESTROY HUMAN COLORECTAL CANCER CELLS

In a study published in 2012 in *Molecular Nutrition & Food Research*, researchers from Taiwan and Boston, Massachusetts, tested the ability of lycopene, a carotenoid that gives fruits and vegetables their red color, and fish oil to destroy human colorectal cancer cells. They began by injecting mice with human colorectal cancer cells. Then, they divided the mice into several groups and fed the mice different combinations of lycopene and/or fish oil. Some mice were put aside to serve as the controls. Both lycopene and fish oil were found to inhibit cancer growth, and the consumption of both together "significantly suppressed" colorectal cancer growth. A low dose of lycopene and fish oil suppressed cancer growth up to 70%; a high dose of the two products suppressed cancer growth up to 85%. The researchers commented that "lycopene and fish oil significantly inhibit the tumor growth, progression, and inflammation in tumor-bearing mice." And, they added that "lycopene and fish oil may potentially act as chemopreventive agents to suppress tumor growth in a mouse xenograft model of CRC [colorectal cancer]."[3]

MAY ACCELERATE CANCER GROWTH

At the same time, in another study published in 2010 in *Cancer Research*, researchers from Michigan (United States) examined the use of fish oil for mice bred to be prone to inflammatory-like bowel disease, a condition that places them at increased risk for colorectal cancer. When the mice were given high doses of fish oil, in just four weeks, they developed late-stage colorectal cancer. Even the mice given the lowest dose of fish oil experienced increases in the severity and aggressiveness of the cancer. The researchers concluded that there may be a need to establish a "tolerable upper limit" for fish oil, especially in people who have chronic conditions such as inflammatory bowel disease.[4]

MAY HELP PREVENT BREAST CANCER

In a study published in 2010 in *Cancer Epidemiology Biomarkers & Prevention*, researchers from Seattle, Washington, and San Diego, California (United

States), examined the association between fish oil intake and breast cancer in a cohort of 35,016 postmenopausal women, ages 50–76, who resided in the western portion of Washington state. After six years, 880 cases of breast cancer were identified. The researchers found that women taking fish oil supplements had reduced risk of developing invasive ductal carcinoma. However, fish oil did not appear to affect their ability to develop another form of breast cancer known as lobular carcinoma. They also learned that fish oil seemed to lower the "risk of local but not regional or distant disease." The researchers suggested that "fish oil may be associated with a reduction of breast cancer risk because of its anti-inflammatory properties." And, they concluded that "fish oil may be inversely associated with breast cancer risk."[5]

MAY IMPROVE NUTRITIONAL STATUS OF PEOPLE RECEIVING CHEMOTHERAPY

It is well known that treatment with chemotherapy may cause severe nausea and extreme fatigue. So, it is not uncommon for the nutritional status of people on chemotherapy to be compromised. They may experience weight loss as well as loss of muscle mass and adipose tissue. Would fish oil be useful for people on chemotherapy? In a study published in 2011 in *Cancer*, researchers from Canada tested fish oil on people receiving chemotherapy for non-small-cell lung cancer. The cohort consisted of 40 patients. During their 10 weeks of chemotherapy, 16 received fish oil and 24 took no supplementation. The researchers found that the subjects on fish oil supplementation were able to maintain their weight. On the other hand, those on no supplementation had an average weight loss of 2.3kg. In addition, those with the highest amounts of serum eicosapentaenoic acid, which is found in fish oil, had the highest amount of gains in muscle mass. Almost 70% of those taking fish oil maintained their prechemotherapy muscle mass or gained mass; less than 30% of the non–fish oil group maintained their muscle mass.

The researchers concluded that their findings "indicate that supplementation with FO [fish oil] ameliorates muscle and adipose tissue wasting in lung cancer patients and provides benefit over patients with SOC [standard of care] receiving first-line chemotherapy."[6]

BUT, MAY BE ILL-ADVISED FOR SOME ON CHEMOTHERAPY

Meanwhile, in an article published in 2011 in *Cancer Cell*, researchers from the Netherlands maintained that fish oil reduces the effectiveness of chemotherapy in mice with tumors under their skin. According to the researchers, under normal conditions, when chemotherapy was administered, the tumors decreased in size. However, when the mice were fed the fatty acids found in fish oil, the tumors were insensitive to the chemotherapy. As a result, the researchers advised people to avoid taking fish oil when undergoing chemotherapy treatments.[7]

MAY INCREASE THE RISK OF
CANCER IN SOME PEOPLE

In a study published in 2012 in the *Archives of Internal Medicine*, researchers based in Paris, France, tested the effects of supplementation with fish oil and B vitamins on more than 2,500 people in France who had a history of heart disease. The researchers divided the subjects into four groups. One group received vitamin B supplementation, another group took fish oil supplementation, a third group took both vitamin B and fish oil supplementation, and the final group took placebos. For almost five years, the researchers tracked the incidences of cancer among the subjects. Over 2,000 people completed the study. During that period, 174 developed cancer, and 58 died from the disease. The researchers found that the men who took fish oil had the same risk of developing cancer as the men who took placebos. However, the women who took fish oil had the three times the risk of developing cancer. And, they were five times more likely to die from cancer.[8]

IS FISH OIL BENEFICIAL FOR CANCER?

There is certainly a good deal of conflicting research on the use of fish oil to prevent and/or treat cancer. Before beginning a fish oil regime, people who may be at increased risk for cancer should discuss fish oil with their medical providers. And, people who have already been diagnosed with cancer should use fish oil with a good deal of caution, if at all.

NOTES

1. Aronson, William J., Naoko Kobayashi, R. James Bernard et al. December 2011. "Phase II Prospective Randomized Trial of a Low-Fat Diet with Fish Oils Supplementation in Men Undergoing Radical Prostatectomy." *Cancer Prevention and Research* 4(12): 2062–2071.

2. Brasky, Theodore M., Cathee Till, Emily White et al. 2011. "Serum Phospholipid Fatty Acids and Prostate Cancer Risk: Results from the Prostate Cancer Prevention Trial." *American Journal of Epidemiology* 173(12): 1429–1439.

3. Tang, F. Y., M. H. Pai, Y. H. Kuo, and X. D. Wang. October 2012. "Concomitant Consumption of Lycopene and Fish Oil Inhibits Tumor Growth and Progression in a Mouse Xenograft Model of Colon Cancer." *Molecular Nutrition & Food Research* 56(10): 1520–1531.

4. Woodworth, Hillary L., Sarah J. McCaskey, David M. Duriancik et al. October 15, 2010. "Dietary Fish Oil Alters T Lymphocyte Cell Populations and Exacerbates Disease in a Mouse Model of Inflammatory Colitis." *Cancer Research* 70(20): 7960–7969.

5. Brasky, Theodore M., Johanna W. Lampe. John D. Potter et al. July 2010. "Specialty Supplements and Breast Cancer Risk in the VITamins And Lifestyle (VITAL) Cohort." *Cancer Epidemiology Biomarkers & Prevention* 19(7): 1696–1708.

6. Murphy, Rachel A., Marina Mourtzakis, Quincy S.C. Chu et al. April 2011. "Nutritional Intervention with Fish Oil Provides a Benefit over Standard of Care for Weight and Skeletal Muscle Mass in Patients with Nonsmall Cell Lung Cancer receiving Chemotherapy." *Cancer* 117(8): 1775–1782.

7. Roodhart, J.M., L.G. Daenen, E.C. Stigter et al. September 13, 2011. "Mesenchymal Stem Cells Induce Resistance to Chemotherapy through the Release of Platinum-Induced Fatty Acids." *Cancer Cell* 20(3): 370–383.

8. Andreeva, V.A., M. Touvier, E. Kesse-Guyot et al. April 9, 2012. "B Vitamin and/or ω-3 Fatty Acid Supplementation and Cancer: Ancillary Findings from the Supplementation with Folate, Vitamins B6 and B12, and/or omega-3 Fatty Acids (SU.FOL.OM3) Randomized Trial." *Archives of Internal Medicine* 172(7): 540–547.

REFERENCES AND RESOURCES

Magazines, Journals, and Newspapers

Andreeva, V.A., M. Touvier, E. Keese-Guyot et al. April 9, 2012. "B Vitamin and/or ω-3 Fatty Acid Supplementation and Cancer: Ancillary Findings from the Supplementation with Folate, Vitamins B6 and B12, and/or Omega-3 Fatty Acids (SU.FOL.OM3) Randomized Trial." *Archives of Internal Medicine* 172(7): 540–547.

Aronson, William J., Naoko Kobayashi, R. James Barnard et al. December 2011. "Phase II Prospective Randomized Trial of a Low-Fat Diet with Fish Oil Supplementation in Men Undergoing Radical Prostatectomy." *Cancer Prevention Research* 4(12): 2062–2071.

Brasky, Theodore M., Cathee Till, Emily White et al. 2011. "Serum Phospholipid Fatty Acids and Prostate Cancer Risk: Results from the Prostate Cancer Prevention Trial." *American Journal of Epidemiology* 173(12): 1429–1439.

Brasky, Theodore M., Joannna W. Lampe, John D. Potter et al. July 2010. "Specialty Supplements and Breast Cancer Risk in the VITamins And Lifestyle (VITAL) Cohort." *Cancer Epidemiology Biomarkers & Prevention* 19(7): 1696–1708.

Murphy, Rachel A., Marina Mourtzakis, Quincy S.C. Chu et al. April 2011. "Nutritional Intervention with Fish Oil Provides a Benefit over Standard of Care for Weight and Skeletal Muscle Mass in Patients with Nonsmall Cell Lung Cancer Receiving Chemotherapy." *Cancer* 117(8): 1775–1782.

Roodhart, J.M., L.G. Daenen, E.C. Stigter et al. September 13, 2011. "Mesenchymal Stem Cells Induce Resistance to Chemotherapy through the Release of Platinum-Induced Fatty Acids." *Cancer Cell* 20(3): 370–383.

Tang, F.Y., M.H. Pai, Y.H. Kuo, and X.D. Wang. October 2012. "Concomitant Consumption of Lycopene and Fish Oil Inhibits Tumor Growth and Progression in a Mouse Xenograft Model of Colon Cancer." *Molecular Nutrition & Food Research* 56(10): 1520–1531.

Woodworth, Hillary L., Sarah J. McCaskey, David M. Duriancik et al. October 15, 2010. "Dietary Fish Oil Alters T Lymphocyte Cell Populations and Exacerbates Disease in a Mouse Model of Inflammatory Colitis." *Cancer Research* 70(20): 7960–7969.

Website

Fred Hutchinson Cancer Research Center. www.fhcrc.org.

Fish Oil and Cardiovascular Health

There have been a number of studies examining fish oil's potential cardiovascular health benefits.

MAY OR MAY NOT SUPPORT
CARDIOVASCULAR HEALTH

In a study published in 2012 in *Arteriosclerosis, Thrombosis, and Vascular Biology*, researchers from Winston-Salem, North Carolina (United States), tested the ability of mega-3 fatty acid oils, such as fish oil, to protect against atherosclerosis, or the build up of plaque in the arteries, in mice bred to be at increased risk for cardiovascular disease. While some mice that were fed regular mouse food served as the controls for 16 days, other mice were fed a higher fat diet that was supplemented with palm oil, echium oil, or fish oil. When compared with the palm oil, the researchers found that both echium oil and fish oil significantly reduced markers of increased cardiovascular risk, such as levels of serum cholesterol. The researchers commented that their data "support the notion that n-3 FAs [omega-3 fatty acids] reduce atherosclerosis."[1]

In a study published in 2009 in *Current Vascular Pharmacology*, researchers from Italy and Turkey conducted an analysis of research studies on the association between intake of fish and fish oil and the control and prevention of high blood pressure. From their examination of the studies on this topic, the researchers determined that the intake of fish and fish oil "could slightly but not significantly reduce systolic and diastolic BP [blood pressure] level and prevent BP increase in either dyslipidaemic, diabetes, elderly, normotensive and hypertensive subjects, contributing to their cardiovascular protective disease protective role."[2]

In a study published in 2005 in *Circulation*, researchers from Boston, Massachusetts (United States), and the Netherlands completed a meta-analysis on randomized, double-blind, placebo-controlled trials on the effects fish oil has on heart rate, a major risk factor for sudden death. After reviewing close to 200 trials, the researchers found 30 that met their inclusion criteria. "In total, this meta-analysis included 1678 individuals treated with fish oil or placebo for 27, 615 person-weeks." The researchers found that fish oil reduced the heart rate of humans, "although the overall effect was modest." The reductions were most dramatic in people with higher baseline heart rates and in people who took fish oil for longer periods of time. The researchers concluded that there is "strong evidence that fish oil consumption directly or indirectly influences cardiac electrophysiology in humans."[3]

Another analysis of trials was published in 2012 in the *British Journal of Nutrition*. This time Spanish researchers examined the association between the consumption of fish oil or blue fish and blood pressure. After analyzing their trials, the researchers found that higher amounts of fish oil and blue fish in the diet "produce a small but significant decrease in blood pressure, especially systolic blood pressure, in older and hypertensive subjects." Unfortunately, "the evidence is not consistent among the different studies." On the other hand, the researchers noted that fish oil and blue fish may "reduce cardiovascular risk in hyperlipemic, diabetic or hypertensive preventing the increase in blood pressure."[4]

Meanwhile, in a randomized, placebo-controlled, double-blind, crossover study published in 2009 in the *International Journal of Food Sciences and Nutrition,* researchers from Japan evaluated the effect that dietary fish oil capsules had on healthy middle-aged Japanese men who ate higher amounts of fish. The cohort consisted of 17 men between the ages of 35 and 64. During the four-week experimental period, the men took either five fish oil or five olive oil (control) capsules with meals. Between treatment phases, there was a four-week washout period. By the end of the study, the researchers found that the fish oil did not result in improvements in markers of cardiovascular health, such as the lowering of total and low-density lipoprotein (LDL or "bad" cholesterol). The researchers wondered if better results would have been obtained if the men normally had lower levels of fish intake.[5]

In a study published in 2012 in the *Journal of Complementary and Integrative Medicine,* researchers from Pennsylvania (United States) wanted to learn if fish oil supplementation would affect the blood pressure of healthy adults. The cohort consisted of 40 adults between the ages of 19 and 55. For six weeks, they were randomly assigned to take fish oil or safflower oil capsules. At the end of the trial, the researchers found that the subjects consuming fish oil had significant decreases in systolic blood pressure (top number). The researchers commented that "the clinical significance of lowering systolic blood pressure in already normotensive adults is unclear."[6]

In a study published in 2009 in the *International Journal of Cardiology,* researchers from Australia conducted a systematic review and meta-analysis of 47 studies on the use of fish oil supplementation for people with elevated lipid levels. These studies included data from a total of 16,511 subjects on levels of cholesterol, high-density lipoprotein (HDL or "good" cholesterol) values, LDL values, and triglyceride readings. Most of the studies were conducted with males, with a mean age of 49, for an average of 24 weeks. The researchers found that fish oil was effective for the reduction of triglyceride levels, but not for reducing total cholesterol, HDL, or LDL. "Fish oil supplementation produced a clinically significant dose-dependent reduction of fasting blood TG [triglyceride levels] but not total, HDL or LDL cholesterol in hyperlipidemic subjects."[7]

In a study that was published in 2012 in *JAMA: The Journal of the American Medical Association,* researchers from Greece reported on their systematic review and meta-analysis of the association between fish oil supplementation and cardiovascular health problems, such as heart attacks and strokes. The 20 studies included in their analysis had 68,680 randomized patients and events that included 7,044 deaths, 3,993 cardiac deaths, 1,150 sudden deaths, 1,837 heart attacks, and 1,490 strokes. The researchers found no significant association between the use of fish oil supplementation and all causes of mortality, cardiac death, sudden death, heart attacks, and stroke. They wrote that fish oils "are not statistically associated with major cardiovascular outcomes across various patient populations." Moreover, they added that their findings provide no justification for the use of fish oils "as a structured intervention in everyday clinical practice."[8]

In a study published in 2012 in *Archives of Internal Medicine*, researchers from Korea conducted a meta-analysis of 14 randomized, double-blind, placebo-controlled trials involving 20,485 patients with a history of cardiovascular disease. The researchers found that fish oil supplementation did not reduce the risk of cardiovascular events such as sudden cardiac death, heart attacks, congestive heart failure, or stoke. And, the researchers observed "insufficient evidence of a secondary effect of secondary preventive effect" of the fish oil "against overall cardiovascular events among patients with a history of cardiovascular disease."[9]

IS FISH OIL BENEFICIAL FOR CARDIOVASCULAR HEALTH?

Obviously, the research on the fish oil benefits for cardiovascular health is far from conclusive. Before beginning supplementation with fish oil to improve cardiovascular health, it is best to have a discussion with your medical provider.

NOTES

1. Brown, A. L., X. Zhu, S. Rong et al. September 2012. "Omega-3 Fatty Acids Ameliorate Atherosclerosis by Favorably Altering Monocyte Subsets and Limiting Monocyte Recruitment to Aortic Lesions." *Arteriosclerosis, Thrombosis, and Vascular Biology* 32(9): 2122–2130.

2. Cicero, Arrigo F.G., Sibel Ertek, and Claudio Borghi. July 2009. "Omega-3 Polyunsaturated Fatty Acids: Their Potential Role in Blood Pressure Prevention and Management." *Current Vascular Pharmacology* 7(3): 330–337.

3. Mozaffarian, Dariush, Anouk Geelen, Ingeborg A. Brouwer et al. 2005. "Effect of Fish Oil on Heart Rate in Humans." *Circulation* 112(13): 1945–1952.

4. Cabo, Jorge, Rodrigo Alonso, and Pedro Mata. June 2012. "Omega-3 Fatty Acids and Blood Pressure." *British Journal of Nutrition* 107(S2): S195–S200.

5. Watanabe, N., Y. Watanabe, M. Kumagai, and K. Fujimoto. September 2009. "Administration of Dietary Fish Oil Capsules in Healthy Middle-Aged Japanese Men with a High Level of Fish Consumption." *International Journal of Food Sciences and Nutrition* 60(S5): 136–142.

6. Noreen, E.E. and J. Brandauer. October 23, 2012. "The Effects of Supplemental Fish Oil on Blood Pressure and Morning Cortisol in Normotensive Adults: A Pilot Study." *Journal of Complementary and Integrative Medicine* 9(1): Article 27.

7. Eslick, Guy D., Peter R. C. Howe, Caroline Smith et al. 2009. "Benefits of Fish Oil Supplementation in Hyperlipidemia: A Systematic Review and Meta-Analysis." *International Journal of Cardiology* 136(1): 4–16.

8. Rizos, E. C., E. E. Ntzani, E. Bika et al. September 12, 2012. "Association between Omega-3 Fatty Acid Supplementation and Risk of Major Cardiovascular Disease Events: A Systematic Review and Meta-Analysis." *JAMA: The Journal of the American Medical Association* 308(10): 1024–1033.

9. Kwak, Sang Mi, Seung-Kwon Myung, Young Jae Lee et al. 2012. "A Meta-Analysis of Randomized, Double-Blind, Placebo-Controlled Trials." *Archives of Internal Medicine* 172(9): 686–694.

REFERENCES AND RESOURCES

Magazines, Journals, and Newspapers

Brown, A. L., X. Zhu, S. Rong et al. September 2012. "Omega-3 Fatty Acids Ameliorate Atherosclerosis by Favorably Altering Monocyte Subsets and Limiting Monocyte Recruitment to Aortic Lesions." *Arteriosclerosis, Thrombosis, and Vascular Biology* 32(9): 2122–2130.

Cabo, Jorge, Rodrigo Alonso, and Pedro Mata. June 2012. "Omega-3 Fatty Acids and Blood Pressure." *British Journal of Nutrition* 107(S2): S195–S200.

Cicero, Arrigo F.G., Sibel Ertek, and Claudio Borghi. July 2009. "Omega-3 Polyunsaturated Fatty Acids: Their Potential Role in Blood Pressure Prevention and Management." *Current Vascular Pharmacology* 7(3): 330–337.

Eslick, Guy D., Peter R. C. Howe, Caroline Smith et al. 2009. "Benefits of Fish Oil Supplementation in Hyperlipidemia: A Systematic Review and Meta-Analysis." *International Journal of Cardiology* 136(1): 4–16.

Kwak, Sang Mi, Seung-Kwon Myung, Young Jae Lee et al. 2012. "Efficacy of Omega-3 Fatty Acid Supplements (Eicosapentaenoic Acid and Docosahexaenoic Acid) in the Secondary Prevention of Cardiovascular Disease: A Meta-Analysis of Randomized, Double-Blind, Placebo-Controlled Trials." *Archives of Internal Medicine* 172(9): 686–694.

Mozaffarian, Dariush, Anouk Geelen, Ingeborg A. Brouwer et al. 2005. "Effect of Fish Oil on Heart Rate in Humans." *Circulation* 112(13): 1945–1952.

Noreen, E. E. and J. Brandauer. October 23, 2012. "The Effects of Supplemental Fish Oil on Blood Pressure and Morning Cortisol in Normotensive Adults: A Pilot Study." *Journal of Complementary and Integrative Medicine* 9(1): Article 27.

Rizos, E. C., E. E. Ntzani, E. Bika et al. September 12, 2012. "Association between Omega-3 Fatty Acid Supplementation and Risk of Major Cardiovascular Disease Events: A Systematic Review and Meta-Analysis." *JAMA: The Journal of the American Medical Association* 308(10): 1024–1033.

Watanabe, N., Y. Watanabe, M. Kumagai, and K, Fujimoto. September 2009. "Administration of Dietary Fish Oil Capsules in Healthy-Middle Aged Japanese Men with a High Level of Fish Consumption." *International Journal of Food Sciences and Nutrition* 60(S5): 136–142.

Website

WebMD. www.webmd.com.

Fish Oil and Children's Health

Pediatricians and family physicians as well as pediatric and family nurse practitioners, who all provide care to young children and adolescents, sometimes wonder if they should advise parents to include fish oil supplements in their children's diets. While it may have some benefits for adults, they wonder if it would improve the health and development of children and adolescents. There are some interesting studies on the topic.

MAY IMPROVE THE OVERALL HEALTH
OF SCHOOLCHILDREN

In a randomized, double-blind, controlled study published in 2012 in the *European Journal of Clinical Nutrition,* researchers from India and the Netherlands wanted to learn if foods fortified with either high or low doses of micronutrients and/or high or low doses of n-3 fatty acids, such as those in fish oil, would affect the morbidity of schoolchildren in India. The researchers divided the children, who were 6–10 years old, into four different parallel intervention groups. One group took high amounts of micronutrients and high amounts of n-3 fatty acids, a second group took low amounts of micronutrients and high amounts of n-3 fatty acids, a third group took high amounts of micronutrients and low amount of n-3 fatty acids, and the fourth group took low amounts of micronutrients and low amounts of n-3 fatty acids. The interventions continued for one year. The researchers found that when they compared the children on lower amounts of n-3 fatty acids, the children consuming higher amounts of n-3 fatty acids had reduced risk for episodes of upper respiratory tract infections, and their infections lasted for shorter periods of time. In addition, they had fewer lower respiratory tract infections, gastrointestinal complaints, and general symptoms of illness, and these problems also had a shorter duration.[1]

In a random, double-blind study published in 2009 in *The Journal of Pediatrics,* researchers from Thailand and the United Kingdom wondered if the consumption of fish oil could keep schoolchildren healthier. The cohort consisted of 180 Thai primary schoolchildren between the ages of 9 and 12. For six months, 94 children took fish oil supplementation and 86 took soybean oil (control) five days per week. The researchers found that the children on fish oil supplementation had fewer episodes of illness, primarily illnesses of the upper respiratory tract, and their illnesses lasted for shorter periods of time.[2]

MAY BE USEFUL FOR ATTENTION-DEFICIT/HYPER
ACTIVITY DISORDER (ADHD)

Researchers from the Yale University School of Medicine in New Haven, Connecticut (United States), conducted a meta-analysis on the use of omega-3 fatty acids, which are found in fish oil, for the symptoms associated with ADHD. The 10 studies included in their meta-analysis had a total of 699 children. The researchers reported on their findings in 2011 in the *Journal of the American Academy of Child and Adolescent Psychiatry*. They noted that they found that "omega-3 fatty acid supplementation demonstrated a small but significant effect in improving ADHD symptoms." This is especially true when the omega-3 fatty acids contain higher amounts of eicosapentaenoic acid (EPA). The researchers commented that the benefits of omega-3 fatty acids was "modest compared with currently available pharmacotherapies." However, it does not have the side effects associated with some medications. And, "it may be reasonable to use omega-3 fatty supplementation to augment traditional pharmacologic interventions or for families who decline other psychopharmacologic options."[3]

In a study published in 2012 in *Nutrition*, researchers from Australia recruited 90 children with ADHD, between the ages of 9 and 12, to participate in a randomized controlled trial. For four months, the children took an EPA-rich fish oil supplement or a docosahexaenoic acid (DHA)-rich fish oil supplement, or a supplement with safflower oil. The researchers found that the children with higher levels of DHA in their red blood cells experienced improvements in word reading and spelling, and their parents noted fewer ADHD symptoms. The improvements were even better for 17 children with learning difficulties. They had significant improvements in word reading, spelling, and the ability to "divide attention." Their parents reported reduced rates of oppositional behavior, hyperactivity, and ADHD symptoms.[4]

In a study published in 2012 in the *Journal of Child Neurology*, researchers from Sri Lanka recruited 94 children between the ages of 6 and 12 who had been diagnosed with ADHD. All of the children had been treated with methylphenidate (Ritalin™), a standard medication for ADHD, and behavior therapy for at least six months. Still, according to their parents, their behavior and academic learning had not improved. The children were randomly assigned to be placed on daily fish oil supplements and evening primrose oil. Though there were no notable changes after three months, after six months parents and teachers reported significant improvements in "inattention, impulsiveness and cooperation." No improvements in distractibility were observed.[5]

MAY PROVIDE PROTECTION FROM ALLERGIES

In a randomized, double-blind, controlled study published in 2012 in *Clinical & Experimental Allergy*, researchers from Australia began with 420 infants considered at high risk for allergies. They were assigned to take daily fish oil supplementation or a control oil from birth until six months of age. At the end of the study, only 120 infants were available for analysis. As may have been expected, the researchers found that the infants taking fish oil supplementation had higher levels of EPA and DHA in their blood. But, they also learned that these infants had significantly lower allergic responses to dust mites and milk protein. The researchers concluded that fish oil supplementation early in life may favorably affect immune patterns and allergy development. And, that fish oil may be "potentially allergy-protective."[6]

MAY BE USEFUL FOR HEALTHY BUT UNDERPERFORMING STUDENTS

In a study published in 2012 in *PLoS ONE*, researchers from the United Kingdom wanted to determine if DHA supplementation would help healthy 7–9-year-old students who were underperforming in reading. Would supplementation with DHA improve reading, working memory, and behavior? The initial cohort consisted of 362 students. For 16 weeks, the students randomly took either DHA capsules or capsules containing corn/soybean oil. The trial was completed by 359 students. The researchers found no overall difference in the reading scores of the two groups. However, when the researchers limited their analysis to the 224 children who had an initial reading ability two years below what was expected for their age, the supplementation significantly improved reading performance. In addition, the supplementation appeared to improve some parent-rated behavior problems. The researchers noted that their trial provided evidence that DHA supplementation "might improve both the behavior and the learning of healthy children from the general school population." The researchers concluded that "dietary deficiencies of DHA might have subtle behavioral effects on children in general."[7]

MAY BE USEFUL FOR CHILDREN AT RISK FOR POOR NUTRITION

In an open label 12-week pilot study published in 2011 in the *Australian and New Zealand Journal of Public Health*, researchers from Australia investigated the use of fish oil supplementation by mostly indigenous children who attended a remote school in the Northern Territory of Australia. All the children, who ranged in age from preschool to grade 7, were invited to participate. Forty-seven children submitted consent forms. The researchers were able to analyze the results from 37 students. They found "anecdotal reports of calmer behavior and better attention during class." Still, the researchers commented that since the study was not placebo-controlled, "it is difficult to attribute the observed changes to supplementation." Yet, they were encouraged by their findings—they "were able to detect significant improvements in achievement relative to age in a cohort of children generally described as underachieving over a relatively short period of fish oil supplementation."[8]

IS FISH OIL BENEFICIAL FOR CHILDHOOD HEALTH?

From the studies reviewed, it is evident that fish oil may well be a good addition to the diets of children. However, before beginning such a regime, it is a good idea to discuss fish oil with your child's medical provider.

NOTES

1. Thomas, T., A. Eilander, S. Muthayya et al. April 2012. "The Effect of a 1-Year Multiple Micronutrient or n-3 Fatty Acid Fortified Food Intervention on Morbidity in Indian School Children." *European Journal of Clinical Nutrition* 66(4): 452–458.

2. Thienprasert, Alice, Suched Samuhaseneetoo, Kathryn Popplestone et al. March 2009. "Fish Oil n-3 Polyunsaturated Fatty Acids Selectively Affect Plasma Cytokines and Decrease Illness in Thai Schoolchildren: A Randomized, Double-Blind, Placebo-Controlled Intervention Trial." *The Journal of Pediatrics* 154(3): 391–395.

3. Bloch, M.H. and A. Qawasmi. October 2011. "Omega-3 Fatty Acid Supplementation for the Treatment of Children with Attention-Deficit/Hyperactivity Disorder Symptomatology: Systematic Review and Meta-Analysis." *Journal of the American Academy of Child and Adolescent Psychiatry* 50(10): 991–1000.

4. Milte, C.M., N. Parletta, J.D. Buckley et al. June 2012. "Eicosapentaenoic and Docosahexaenoic Acids, Cognition, and Behavior in Children with Attention-Deficit/Hyperactivity Disorder: A Randomized Controlled Trial." *Nutrition* 28(6): 670–677.

5. Perera, H., K.C. Jeewandara, S. Seneviratne, and C. Guruge. June 2012. "Combined ω3 and ω6 Supplementation in Children with Attention-Deficit Hyperactivity Disorder (ADHD) Refractory to Methylphenidate Treatment: A Double-Blind, Placebo-Controlled Study." *Journal of Child Neurology* 27(6): 747–753.

6. D'Vaz, N., S.J. Meldrum, J.A. Dunstan et al. August 2012. "Fish Oil Supplementation in Early Infancy Modulates Developing Infant Immune Responses." *Clinical & Experimental Allergy* 42(8): 1206–1216.

7. Richardson, A.J., J.R. Burton, R.P. Sewell et al. 2012. "Docosahexaenoic Acid for Reading, Cognition and Behavior in Children Aged 7–9 Years: A Randomized Controlled Trial (the DOLAB Study)." *PLoS ONE* 7(9): e43909.

8. Sinn, Natalie, Patrick Cooper, and Kerin O'Dea. October 2011. "Fish Oil Supplementation, Learning and Behaviour in Indigenous Australian Children from a Remote Community School: A Pilot Feasibility Study." *Australian and New Zealand Journal of Public Health* 35(5): 493–494.

REFERENCES AND RESOURCES

Magazines, Journals, and Newspapers

Bloch, M.H. and A. Qawasmi. October 2011. "Omega-3 Fatty Acid Supplementation for the Treatment of Children with Attention-Deficit/Hyperactivity Disorder Symptomatology: Systematic Review and Meta-Analysis." *Journal of the American Academy of Child and Adolescent Psychiatry* 50(10): 991–1000.

D'Vaz, N., S.J. Meldrum, J.A. Dunstan et al. August 2012. "Fish Oil Supplementation in Early Infancy Modulates Developing Infant Immune Responses." *Clinical & Experimental Allergy* 42(8): 1206–1216.

Milte, C.M., N. Parletta, J.D. Buckley et al. June 2012. "Eicosapentaenoic and Docosahexaenoic Acids, Cognition, and Behavior in Children with Attention-Deficit/Hyperactivity Disorder: A Randomized Controlled Trial." *Nutrition* 28(6): 670–677.

Perera, H., K.C. Jeewandara, S. Seneviratne, and C. Guruge. June 2012. "Combined ω3 and ω6 Supplementation in Children with Attention-Deficit Hyperactivity

Disorder (ADHD) Refractory to Methylphenidate Treatment: A Double-Blind, Placebo-Controlled Study." *Journal of Child Neurology* 27(6): 747–753.

Richardson, A. J., J. R. Burton, R. P. Sewell et al. 2012. "Docosahexaenoic Acid for Reading, Cognition and Behavior in Children Aged 7–9 Years: A Randomized, Controlled Trial (the DOLAB Study)." *PLoS ONE* 7(9): e43909.

Sinn, Natalie, Patrick Cooper, and Kerin O'Dea. October 2011. "Fish Oil Supplementation, Learning and Behaviour in Indigenous Australian Children from a Remote Community School: A Pilot Feasibility Study." *Australian and New Zealand Journal of Public Health* 35(5): 493–494.

Thienprasert, Alice, Suched Samuhaseneetoo, Kathryn Popplestone et al. March 2009. "Fish Oil n-3 Polyunsaturated Fatty Acids Selectively Affect Plasma Cytokines, and Decrease Illness in Thai Schoolchildren: A Randomized, Double-Blind, Placebo-Controlled Intervention Trial." *The Journal of Pediatrics* 154(3): 391–395.

Thomas, T., A. Eilander, S. Muthayya et al. April 2012. "The Effect of a 1-Year Multiple Micronutrient or n-3 Fatty Acid Fortified Food Intervention on Morbidity in Indian School Children." *European Journal of Clinical Nutrition* 66 (4): 452–458.

Website

The Royal Children's Hospital Melbourne. www.rch.org.

Fish Oil and Pregnancy

Some contend that pregnant women and their soon to be born infant benefit from supplementation with fish oil. Others maintain that such supplementation is not necessary. And, still others state that fish oil supplementation has the potential to have short- and long-term detrimental effects on both the mother and child.

MAY OR MAY NOT BE A USEFUL ADDITION TO THE PRENATAL DIET

In a study published in 2011 in *Acta Obstetricia et Gynecologica Scandinavica*, researchers from Sweden and Detroit, Michigan (United States), noted that the delivery of infants before the 37th week of gestation is not as uncommon as some believe. In fact, in the developed world, the early delivery of babies represents about 5%–10% of deliveries. The researchers wondered if the increased consumption of fish oil during pregnancy would lower the number of preterm births. So, they reviewed research from several studies, including a total of 921 women for whom there was information on gestational ages and 1,187 for whom there was information on birth weights. The researchers found that women who had larger amounts of fish oil in their diets had fewer preterm deliveries, and their babies had higher birth weights. They concluded that fish oil "may delay the timing of spontaneous delivery and be beneficial in relation to prevention of PTB [pre-term birth] and its associated complications."[1]

In a randomized, double-blind, controlled trial published in 2011 in *Pediatrics*, re-searchers from Atlanta, Georgia (United States), and Mexico wanted to determine if the use of docosahexaenoic acid (DHA), found in fish oil, during pregnancy influ-enced the morbidity of the infant. The researchers recruited the women in Mexico; they were all between the ages of 18 and 35 and were in gestation weeks 18–22. The 1,094 women who were in the initial cohort were assigned to receive either DHA or a placebo with corn and soy oils, which they took until delivery. Eighty-nine per-cent of the women completed the trial. And, morbidity data were available for 849, 834, and 834 infants at one, three, and six months, respectively. When compared to the infants whose mothers were controls, at one month of age, the infants of the mothers who took the supplementation had fewer cold symptoms. In addition, the supplementation "influenced illness symptoms duration at 1, 3, and 6 months."[2]

In a systematic review of research on the intake of long-chain polyun-saturated fatty acids during pregnancy published in 2012 in *Paediatric and Perinatal Epidemiology*, researchers from Atlanta, Georgia, and Worcester, Mas-sachusetts (both United States), evaluated 15 randomized controlled trials and 14 observational reports. They found that the infants of the women on supple-mentation had "modest" increases in birth weight but no significant differences in birth length or head circumference. The women on supplementation had a lower risk of giving birth before the 34th week of gestation, and they appeared to have a lower risk of preterm delivery and a reduced risk of having an infant with a low birth weight. Still, the researchers noted that it is not uncommon for preg-nant women to have low intakes of long-chain polyunsaturated fatty acids. "Preg-nant women in many low, middle, and high income countries do not achieve the recommended intake of at least 200mg of DHA per day." This is especially true in poor countries such as India and Bangladesh.[3]

In a study published in 2011 in *The American Journal of Clinical Nutrition*, researchers from Germany wanted to learn if the amount of fatty acids in the umbilical cord serum had any effects on a child's behavior years later, when the child was 10. Their cohort consisted of 416 children. The researchers found that the children who had higher concentrations of DHA in their cord blood serum had reduced rates of hyperactivity and inattention. The researchers noted that they found a "strong association" between the amount of DHA in the core blood "and lower scores on overall behavior difficulties in children at 10 years of age." This association exists "independent of the family's socioeconomic status, ad-verse factors in pregnancy, or actual dietary intake." The researchers concluded that "perinatal DHA availability may play a critical role in the pathogenesis of a clinically significant behavior problem."[4]

In a trial published in 2012 in *BMJ*, researchers from Australia wondered if fish oil supplementation during pregnancy would have any effect on unborn ba-bies that have a higher risk of becoming allergic infants. (The unborn babies were thought to be a higher risk if their mother, father, or sibling had a history of a medically diagnosed allergic disease.) The cohort consisted of 368 women who were randomly allocated to receive fish oil capsules and 338 women who received vegetable oil capsules (rapeseed, sunflower, and palm oils) beginning the 21th week of gestation. They continued the supplementation until all of their 706 babies were born. The researchers found no difference in the incidence of eczema or food allergies between the babies in the supplemental and control groups, although the babies from the supplement group had fewer cases of infant atopic eczema and fewer infants demonstrated a sensitivity to eggs.[5]

In a study published in 2011 in *The American Journal of Clinical Nutrition*, researchers from Australia wanted to learn if the intake of fish oil by a pregnant woman would improve the visual acuity of infants who were four months old. The researchers examined the results of research that was conducted on pregnant women enrolled in a randomized, double-blind trial. From mid-pregnancy until delivery, these women were assigned to consume fish oil or vegetable oil capsules. The researchers then conducted tests to determine the visual acuity of 185 infants. The researchers found no differences in the visual acuity of the infants of the mothers on either type of oil. They concluded that fish oil supplementation during pregnancy "does not enhance infant visual acuity in infants at 4 months of age."[6]

In a study published in 2011 in *Lipids*, researchers from Denmark wanted to determine if supplementation during the third trimester of pregnancy would have any effect, almost two decades later, on the lipid or lipoprotein levels of the offspring. The study was a follow-up of a previous randomized, controlled trial in which 533 women in their 30th week of pregnancy were placed on fish oil ($n = 266$) or olive oil ($n = 136$) or no oil ($n = 131$). In 2009, the researchers invited the offspring to have a physical examination and blood testing. Two hundred and forty-three of the offspring, who were between 18 and 19, participated. The researchers found no association between fish oil supplementation during the last trimester of pregnancy and the lipid profiles of the offspring.[7]

In a study published in 2011 in *The American Journal of Clinical Nutrition*, the same group of researchers from Denmark examined the same subjects to determine if there was an association of intake of fish oil during the last trimester of pregnancy and adiposity (fat) in bodies of the 18–19-year-old offspring. The researchers found no difference in the body mass index (BMI) between the children of the pregnant women who took fish oil or olive oil supplementation. "Overall, results of the biochemical analyses supported the finding of no difference between the groups." As a result, supplementation with fish oil during the third trimester of pregnancy is "not associated with adiposity" in the offspring.[8]

IS FISH OIL A BENEFICIAL
PRENATAL SUPPLEMENT?

From the research reviewed, it is almost impossible to know if pregnant women should be taking fish oil supplementation. Before a pregnant woman begins taking fish oil supplement, she should discuss the possibility with her medical provider.

NOTES

1. Salvig, J.D. and R.F. Lamont. August 2011. "Evidence Regarding an Effect of Marine n-3 Fatty Acids on Preterm Birth: A Systematic Review and Meta-Analysis." *Acta Obstetricia et Gynecologica Scandinavica* 90(8): 825–838.

2. Imhoff-Kunsch, B., A. D. Stein, R. Martorll et al. September 2011. "Prenatal Docosahexaenoic Acid Supplementation and Infant Morbidity: Randomized Controlled Trial." *Pediatrics* 128(3): e505–e512.

3. Imhoff-Kunsch, B., V. Briggs, T. Goldenberg, and U. Ramakrishnan. July 2012. "Effect of n-3 Long-Chain Polyunsaturated Fatty Acid Intake during Pregnancy on Maternal, Infant, and Child Health Outcomes: A Systematic Review." *Paediatric and Perinatal Epidemiology* 26(s1): 91–107.

4. Kohlboeck, G., C. Glaser, C. Tiesler et al. December 2011. "Effect of Fatty Acid Status in Cord Blood Serum on Children's Behavioral Difficulties at 10 Years of Age: Results from the LISAplus Study." *The American Journal of Clinical Nutrition* 94(6): 1592–1599.

5. Palmer, D.J., T. Sullivan, M.S. Gold et al. 2012. "Effect of n-3 Long Chain Polyunsaturated Fatty Acid Supplementation in Pregnancy on Infants' Allergies in the First Year of Life: Randomised Controlled Trial." *BMJ* 344: e184.

6. Smithers, L.G., R.A. Gibson, and M. Makrides. June 2011. "Maternal Supplementation with Docosahexaenoic Acid during Pregnancy Does Not Affect Early Visual Development in the Infant: A Randomized Controlled Trial." *The American Journal of Clinical Nutrition* 93(6): 1293–1299.

7. Rytter, Dorte, Erik B. Schmidt, Bodil H. Bech et al. December 2011. "Fish Oil Supplementation during Late Pregnancy Does Not Influence Plasma Lipids or Lipoprotein Levels in Young Adult Offspring." *Lipids* 46(12): 1091–1099.

8. Rytter, Dorte, Bodil H. Bech, Jeppe H. Christensen et al. September 2011. "Intake of Fish Oil during Pregnancy and Adiposity in 19-Year-Old Offspring: Follow-Up on a randomized Controlled Trial." *The American Journal of Clinical Nutrition* 94(3): 701–708.

REFERENCES AND RESOURCES

Magazines, Journals, and Newspapers

Imhoff-Kunsch, B., A.D. Stein, R. Martorell et al. September 2011. "Prenatal Docosahexaenoic Acid Supplementation and Infant Morbidity: Randomized Controlled Trial." *Pediatrics* 128(3): e505–e512.

Imhoff-Kunsch, B., V. Briggs, T. Goldenberg, and U. Ramakrishnan. July 2012. "Effect of n-3 Long-Chain Polyunsaturated Fatty Acid Intake during Pregnancy on Maternal, Infant, and Child Health Outcomes: A Systematic Review." *Paediatric and Perinatal Epidemiology* 26(s1): 91–107.

Kohlboeck, G., C. Glaser, C. Tiesler at al. December 2011. "Effect of Fatty Acid Status in Cord Blood Serum on Children's Behavioral Difficulties at 10 Years of Age: Results from the LISAplus Study." *The American Journal of Clinical Nutrition* 94(6): 1592–1599.

Palmer, D.J., T. Sullivan, M.S. Gold et al. January 30, 2012. "Effect of n-3 Long Chain Polyunsaturated Fatty Acid Supplementation in Pregnancy on Infants' Allergies in First Year of Life: Randomised Controlled Trial." *BMJ* 344: e184.

Rytter, Dorte, Bodil H. Bech, Jeppe H. Christensen et al. September 2011. "Intake of Fish Oil during Pregnancy and Adiposity in 19-Year-Old Offspring: Follow-Up on a Randomized Controlled Trial." *The American Journal of Clinical Nutrition* 94(3): 701–708.

Rytter, Dorte, Erik B. Schmidt, Bodil H. Bech et al. December 2011. "Fish Oil Supplementation during Late Pregnancy Does Not Influence Plasma Lipids or Lipoprotein Levels in Young Adult Offspring." *Lipids* 46(12): 1091–1099.

Salvig, J. D. and R. F. Lamont. August 2011. "Evidence Regarding an Effect of Marine n-3 Fatty Acids on Preterm Birth: A Systematic Review and Meta-Analysis." *Acta Obstetricia et Gynecologica Scandinavica* 90(8): 825–838.

Smithers, L. G., R. A. Gibson, and M. Makrides. June 2011. "Maternal Supplementation with Docosahexaenoic Acid during Pregnancy Does Not Affect Early Visual Development in the Infant: A Randomized Controlled Trial." *The American Journal of Clinical Nutrition* 93(6): 1293–1299.

Website

American Pregnancy Association. www.americanpregnancy.org.

Fish Oil and Psychiatric Health

It has been said that fish oil is useful for psychiatric conditions such as anxiety and depression. Would the millions of people who suffer every day from these medical problems benefit from fish oil? If helpful, it would certainly be useful for those debilitated by these illnesses.

MAY OR MAY NOT BE USEFUL FOR PSYCHIATRIC CONDITIONS

In a randomized, double-blind, controlled study published in 2009 in *The Journal of Clinical Psychiatry*, researchers from Boston, Massachusetts (United States), compared the treatment of subjects with a major depressive disorder with ethyl-eicosapentaenoate (EPA-E), found in fish oil supplements and cold-water fish, to the treatment with a placebo. For eight weeks, 57 subjects took either 1g/day of EPA or a placebo. Twenty-four subjects completed the entire trial. The researchers found that the treatment did, indeed, relieve depressive symptoms. "EPA appears to be a well-tolerated, potentially effective monotherapy for MDD [major depressive disorder] at doses of 1g/day." Still, the researchers noted that their results "must be interpreted with caution, in view of the small sample and modest number of completers, which limited statistical power."[1]

In a study published in 2011 in *Behavioural Brain Research*, researchers from Brazil investigated the use of fish oil on rats subjected to a number of different behaviors that induced anxiety and depression. The researchers first divided the rats into four groups—control rats, control rats subjected to stress, rats fed fish oil but not subjected to stress, and rats fed fish oil and subjected to stress. The researchers found that the rats fed supplemental fish oil from a relatively early age were better able to deal with stress-provoking activities. Fish oil "prevented the occurrence of anxiety like behaviors, depressive-like behaviors and deficits of learning and memory in rats subjected to restraint stress."[2]

In a randomized, double-blind, placebo-controlled study published in 2011 in the *European Archives of Psychiatry and Clinical Neuroscience*, researchers from Iran and Dallas, Texas (United States), wanted to determine if fish oil would help elders with mild to moderate depression. The initial cohort consisted of 66 people who were 65 or older. For six months, half of the subjects received daily supplementation of 1g of fish oil; the other half took placebo supplementation. The researchers found that the fish oil "had some efficacy in the treatment of mild to moderate depression in elderly subjects."[3]

In a randomized, placebo-controlled, parallel group study published in 2011 in *Brain, Behavior, and Immunity*, researchers from Ohio (United States) wanted to learn if supplementation with EPA and docosahexaenoic acid (DHA), which are found in fish oil, was useful for medical students dealing with anxiety during periods of higher and lower levels of stress. The cohort consisted of 38 men and 30 women, who were first- and second-year medical students, between the ages of 21 and 29. For 12 weeks, the students took either the EPA/DHA supplement or a placebo containing palm, olive, canola, and coco butter oils "that approximated the saturated:monounsaturated:polyunsaturated . . . ratio consumed by US adults." When compared to the controls, students who took the supplements had 20% less anxiety. The supplements did not appear to have any effect on depression. According to the researchers, this is the first trial that shows reductions in anxiety from supplemental EPA/DHA in individuals without an anxiety disorder diagnosis.[4]

In a randomized, double-blind study published in 2007 in *Progress in Neuro-Psychopharmacology & Biological Psychiatry*, researchers from Australia wanted to learn if supplementation with tuna fish oil would help people dealing with major depression. The cohort consisted of 83 subjects between the ages of 18 and 70. When the trial began, 61 subjects were already taking therapeutic doses of antidepressants. All of the subjects were assigned to take eight odorless 1g soft gelatin capsules per day of either South Pacific tuna oil or olive oil (placebo). At the end of 16 weeks, there were large reductions in depression, but they were not statistically significant. The researchers concluded that their "findings do not support the routine use of this particular tuna fish oil as an addition to conventional treatment for major depression."[5]

In a study published in 2012 in *The Journal of Clinical Psychiatry*, researchers from Australia conducted a meta-analysis to determine if omega-3 fatty acids, such as fish oil, were useful for mania and bipolar depression. Although the researchers began with 168 studies, only 6 randomized, controlled trials met their criteria for inclusion. Their six studies had a mean trial length of 12.6 weeks and a mean sample size of 37.8. The researchers found that omega-3 supplementation "significantly reduced depressive symptoms over control interventions." According to the researchers, "meta-analytic evidence strongly supports the use of omega-3 for adjuvant use in the treatment of depressed mood in bipolar disorder." At the same time, omega-3 supplementation does not appear to be useful for mania.[6]

In a study published in 2008 in *The American Journal of Clinical Nutrition*, different researchers from the Netherlands investigated the intake of fish oil and mental well-being in older persons who live independently. The cohort consisted of 302 people who were at least 65; the mean age was 70. All the subjects were assigned to take a higher (1,800mg—equivalent to eating eight portions of fish per week) or lower (400mg—equivalent to eating two portions of fish per week, one of which is oily fish) dose of fish oil supplement or a placebo. By the end of the trial, there were 299 participants. The researchers found "no effect of daily supplementation with high or low doses" of fish oil on "mental well-being as assessed by depression and anxiety questionnaires."[7]

In a randomized, double-blind study published in 2012 in the *Journal of Psychopharmacology*, researchers from the Netherlands examined the effects of fish oil supplementation on cognition and mood of 71 people who recovered from at least one major depressive episode. Since one participant left the trial, the analysis was calculated based on results from the remaining 70 subjects. The researchers found that the subjects taking fish oil experienced decreases in tension and depression scores. Fish oil had only a marginal effect on fatigue. And, no significant effects were seen in "attention, cognitive reactivity, and depressive symptoms." Though the researchers acknowledged that their findings were "inconclusive," they noted that their results may indicate that fish oil "supplementation has selective effects on emotional cognition and mood in recovered depressed participants."[8]

In a trial published in 2010 in *JAMA: The Journal of the American Medical Association*, researchers from Australia hypothesized that increasing the intake of fish oil during the last half of pregnancy would result in fewer women with depressive symptoms and enhance the neurodevelopmental outcome of the children. The initial cohort consisted of 2,399 women who were pregnant with a single child. The women, who were enrolled before they reached 21 weeks of gestation, were randomly assigned to take three fish oil or vegetable oil tablets per day. They continued the supplementation until their babies were born. The researchers were able to collect sufficient data on 2,320 women. In addition, they assessed 694 children at 18 months. The researchers found no significant difference in depressive symptoms between the women taking fish oil and the women taking vegetable oil. Likewise, there was no significant difference in the cognitive scores of the children. The researchers concluded that when compared with vegetable oil capsules, the fish oil capsules "did not result in lower levels of postpartum depression in mothers or improved cognitive and language development in offspring during early childhood."[9]

IS FISH OIL BENEFICIAL FOR
PSYCHIATRIC PROBLEMS?

It is not clear. Fish oil supplementation may well be useful for some people dealing with psychiatric problems. But, the results of the studies reviewed are mixed. So, before beginning a fish oil regime, it is probably a good idea to discuss it with your medical provider.

NOTES

1. Mischoulon, D., G. Papakostas, C. Dording et al. December 2009. "A Double-Blind Randomized Controlled Trial of Ethyl-Eicosapentaenoate [EPA-E] for Major Depressive Disorder." *The Journal of Clinical Psychiatry* 70(12): 1636–1644.

2. Ferraz, A.C., A.M. Delattre, R.G. Almendra et al. May 16, 2011. "Chronic ω-3 Fatty Acids Supplementation Promotes Beneficial Effects on Anxiety, Cognition and Depressive-Like Behaviors in Rats Subjected to a Restraint Stress Protocol." *Behavioural Bran Research* 219(1): 116–122.

3. Tajalizadekhoob, Y., F. Sharifi, H. Fakhrzadeh et al. December 2011. "The Effect of Low-Dose Omega 3 Fatty Acids on the Treatment of Mild to Moderate Depression in the Elderly: A Double-Blind, Randomized, Placebo-Controlled Study." *European Archives of Psychiatry and Clinical Neuroscience* 261(8): 539–549.

4. Kiecolt-Glaser, Janice, Martha A. Belury, Rebecca Andridge et al. November 2011. "Omega-3 Supplementation Lowers Inflammation and Anxiety in Medical Students: A Randomized Controlled Trial." *Brain, Behavior, and Immunity* 25(8): 1725–1734.

5. Grenyer, Brin F.S., Trevor Crowe, Barbara Meyer et al. 2007. "Fish Oil Supplementation in the Treatment of Major Depression: A Randomised Double-Blind Placebo-Controlled Trial." *Progress in Neuro-Psychopharmacology & Biological Psychiatry* 31(7): 1393–1396.

6. Sarris, Jerome, David Mischoulon, and Isaac Schweitzer. January 2012. "Omega-3 for Bipolar Disorder: Meta-Analysis of Use in Mania and Bipolar Depression." *The Journal of Clinical Psychiatry* 73(1): 81–86.

7. van de Rest, Ondine, Johanna M. Geleijnse, Frans J. Kok et al. September 2008. "Effect of Fish-Oil Supplementation on Mental Well-Being in Older Subjects: A Randomized, Double-Blind, Placebo-Controlled Trial." *The American Journal of Clinical Nutrition* 88(3): 706–713.

8. Antypa, Niki, August H.M. Smelt, Annette Strengholt, and A.J. Willem van der Does. May 2012. "Effects of Omega-3 Fatty Acid Supplementation on Mood and Emotional Information Processing in Recovered Depressed Individuals." *Journal of Psychopharmacology* 26(5): 738–743.

9. Makrides, Maria, Robert A. Gibson, Andrew J. McPhee et al. October 20, 2010. "Effect of DHA Supplementation during Pregnancy on Maternal Depression and Neurodevelopment of Young Children: A Randomized, Controlled Trial." *JAMA: The Journal of the American Medical Association* 304(15): 1675–1683.

REFERENCES AND RESOURCES
Magazines, Journals, and Newspapers

Antypa, Niki, August H.M. Smelt, Annette Strengholt, and A.J. Willem van der Does. May 2012. "Effects of Omega-3 Fatty Acid Supplementation on Mood and Emotional Information Processing in Recovered Depressed Individuals." *Journal of Psychopharmacology* 26(5): 738–743.

Ferraz, A.C., A.M. Delattre, R.G. Almendra et al. May 16, 2011. "Chronic ω-3 Fatty Acids Supplementation Promotes Beneficial Effects on Anxiety, Cognition and Depressive-Like Behaviors in Rats Subjected to a Restraint Stress Protocol." *Behavioural Brain Research* 219(1): 116–122.

Grenyer, Brin F.S., Trevor Crowe, Barbara Meyer et al. 2007. "Fish Oil Supplementation in the Treatment of Major Depression: A Randomised Double-Blind Placebo-Controlled Trial." *Progress in Neuro-Psychopharmacology & Biological Psychiatry* 31(7): 1393–1396.

Kiecolt-Glaser, Janice, Martha A. Belury, Rebecca Andridge et al. November 2011. "Omega-3 Supplementation Lowers Inflammation and Anxiety in Medical Students: A Randomized Controlled Trial." *Brain, Behavior, and Immunity* 25(8): 1725–1734.

Makrides, Maria, Robert A. Gibson, Andrew J. McPhee et al. October 20. 2010. "Effect of DHA Supplementation during Pregnancy on Maternal Depression and Neurodevelopment of Young Children: A Randomized Controlled trial." *JAMA: The Journal of the American Medical Association* 304(15): 1675–1683.

Mischoulon, D., G. Papakostas, C. Dording et al. December 2009. "A Double-Blind Randomized Controlled Trial of Ethyl-Eicosapentaenoate (EPA-E) for Major Depressive Disorder." *The Journal of Clinical Psychiatry* 70(12): 1636–1644.

Sarris, Jerome, David Mishoulon, and Isaac Schweitzer. January 2012. "Omega-3 for Bipolar Disorder: Meta-Analysis of Use in Mania and Bipolar Depression." *The Journal of Clinical Psychiatry* 73(1): 81–86.

Tajalizadekhoob, Y., F. Sharifi, and H. Fakhrzadeh el al. December 2011. "The Effect of Low-Dose Omega 3 Fatty Acids on the Treatment of Mild to Moderate Depression in the Elderly: A Double-Blind, Randomized, Placebo-Controlled Study." *European Archives of Psychiatry and Clinical Neuroscience* 261(8): 539–549.

van de Rest, Ondine, Johanna M. Geleijnse, Frans J. Kok et al. September 2008. "Effect of Fish-Oil Supplementation on Mental Well-Being in Older Subjects: A Randomized, Double-Blind, Placebo-Controlled Trial." *The American Journal of Clinical Nutrition* 88(3): 706–713.

Website

WebMD. www.webmd.com.

Flaxseed Oil: Overview

Flaxseeds are obtained from the flax plant, and flaxseed oil is made from flaxseeds. Although the flax plant is thought to have first grown in Egypt, it now cultivated throughout Canada and the Northwestern sections of the United States.[1]

Flaxseed oil contains both omega-3 and omega-6 fatty acids. It also has the essential fatty acid alpha-linolenic acid, which the body converts into eicosapentaenoic acid and docosahexaenoic acid.[2] Flaxseed oil is believed to support cardiovascular health and prevent cancer. Furthermore, since it is thought to have laxative properties, it may be useful for constipation. Many women contend that flaxseed oil decreases hot flashes. And, it is said to help with the symptoms

Flax flowers in bloom in a field. (Elena
Elisseeva/Dreamstime.com)

of Sjögren's syndrome, an autoimmune disorder in which the body attacks the
glands that produce moisture, such as the salivary and tear glands.

Flaxseed oil is sold in conventional stores, specialty stores, and online. It is
relatively inexpensive.

MAY BE USEFUL FOR SKIN CONDITIONS

In a randomized, double-blind, placebo-controlled trial published in 2009 in
the British Journal of Nutrition, researchers from Germany and France tested the
ability of flaxseed and borage oils to help women with sensitive and dry skin. The
initial cohort consisted of 45 healthy, nonsmoking women between the ages of
18 and 65. The women were divided into three groups of 15. One group took supplemental flaxseed oil, one group took supplemental borage oil, and the women
in the third group took a placebo. By the end of the 12-week trial, both flaxseed
oil and borage oil had significantly increased skin hydration. Both groups had reductions in transepidermal water loss and reductions in roughness and skin scaling. The researchers concluded that their findings indicated "that skin properties
can be modulated by an intervention with dietary lipids."[3]

In a study published in 2011 in *Skin Pharmacology and Physiology*, most of the previous researchers from Germany and France compared supplementation with flaxseed oil to supplementation with safflower seed oil. The cohort, which consisted of 26 nonsmoking women between the ages of 18 and 65, was divided into two groups of 13. For 12 weeks, the women were randomly assigned to take flaxseed oil supplementation or safflower seed oil supplementation. The researchers found that supplementation with flaxseed oil resulted in significant reductions in skin sensitivity, transepidermal water loss, skin roughness, and scaling. At the same time, the skin was smoother and better hydrated. Meanwhile, the women taking the safflower seed oil supplementation had "only a significant improvement in skin roughness and hydration."[4]

MAY BE USEFUL FOR REPAIRING WOUNDS

In a study published in 2012 in *Evidence-Based Complementary and Alternative Medicine*, researchers from Brazil investigated the use of flaxseed oil to treat skin wounds on rats. The cohort consisted of 72 male and female Wistar rats that were randomly divided into groups. After wounds were administered to the rats, some rats were set aside to serve as controls. The wounds of the remaining rats were treated with different doses of regular flaxseed oil or a semisolid formulation of flaxseed oil. The researchers found that the topical administration of the semisolid 1% or 5% formulation of flaxseed oil promoted healing in all the animals treated, "therefore indicating the potential for therapeutic action . . . when used at low concentrations in the preparation of dermatological formulas with a solid base."[5]

MAY PROTECT SKIN FROM DAMAGE

In a 28-day study published in 2012 in *Toxicology and Industrial Health*, researchers from Turkey wanted to learn if flaxseed oil would protect the skin of rats exposed to ultraviolet C (UVC)—the highest energy and most dangerous form of ultraviolet light. The researchers began by dividing 21 Sprague-Dawley male albino rats into three groups. The rats in the first group served as the controls. The rats in the second group were exposed to UVC light for 1 hour twice a day; the rats in the third groups were exposed to the same amount of UVC light, but their diets were supplemented with flaxseed oil. The researchers found that the UVC light damaged the antioxidant defense system of the rats and induced cell death (apoptosis) in various tissues. However, flaxseed supplementation prevented some of "the detrimental effects of UVC light or at least . . . it reduced the injury."[6]

MAY OFFER SOME HELP FOR DRY EYES

In a pilot, prospective, randomized, double-masked study published in 2011 in *Cornea*, researchers based in Texas examined the ability of a supplement that contained flaxseed oil and fish oil to help people who suffered from a medical problem causing their eyes to be chronically dry. The cohort consisted of 36 subjects; 21 were in the treatment group and 15 were in the placebo group. For 90 days, they took either a daily supplement containing flaxseed oil and fish oil or a placebo (wheat germ oil). The results were mixed. Treatment with the combined supplement resulted in increased tear production and tear volume. But other problems, such as the tear evaporation rate, did not improve.[7]

MAY HAVE ANTICANCER PROPERTIES

In a study published in 2011 in the *Asian Pacific Journal of Cancer Prevention*, researchers based in Egypt investigated the ability of flaxseed oil to prevent colon cancer in rats. The researchers began by dividing male Wistar rats into six groups. The rats in groups 1, 3, and 5 were given a chemical to induce colon cancer; the rats in groups 1 and 3 were fed a diet containing 20% or 5% flaxseed oil. The rats in groups 2 and 4 served as flaxseed dose corresponding controls, but they were not treated with the chemical to induce colon cancer. The rats in the sixth group were negative controls. After 32 weeks, the animals were sacrificed. The researchers found "clear inhibitory effects of dietary crude Egyptian flaxseed oil on post-initiation stages of rat colon carcinogenesis and colon tumors." Moreover, the researches hypothesized that the "antiproliferative properties of ω-3 polyunsaturated fatty acid found in α-linolenic acid and flax lignans both found in flaxseeds oils."[8]

MAY HELP PREVENT SEIZURES AND DEPRESSION

In a study published in 2012 in *Pharmacological Research*, researchers from India wondered if flaxseed oil or the Ayurvedic formulation *Ashwagandharishta* or the combination of these two treatments may be useful in the prevention of seizures, and the depression that is often associated with seizures. The researchers began by noting that there are a number of medications for seizures. But, they have negative side effects. In addition, "one patient out of three is resistant to antiepileptic drugs." So, the researchers tested the two treatments separately and together in albino rats that had electroshock-induced seizures. Two groups served as controls, and the rats in one group were treated with phenytoin sodium, a medication for epilepsy. The researchers found both products to be useful for seizures and depression. "Both *Ashwagandharishta* and flaxseed oil have antiepileptic activity; besides, they are having excellent anti-post-ictal [post-seizure] depression." Both of these treatments "can play a major role as an adjuvant therapy with modern antiepileptic drugs."[9]

MAY HELP PROTECT THE LIVER FROM
CISPLATIN, A POTENT CHEMOTHERAPY

In a study published in 2012 in *Human & Experimental Toxicology*, researchers from India noted that cisplatin is a very effective treatment for a variety of solid tumors. However, when used in higher amounts, it may be toxic to the liver, a condition known as hepatotoxicity. Could flaxseed oil provide a degree of protection to the liver? The researchers decided to test this theory in adult male Wistar rats, which they divided into four groups. During the first part of the study, the rats were fed either a normal diet or a diet containing 15% flaxseed oil. After 10 days, the rats in two of the groups were administered a single dose of cisplatin. Four days later, the rats were sacrificed. The researchers found that the deleterious effects of cisplatin on the liver were ameliorated by flaxseed oil. And, they concluded that supplementation with flaxseed oil may enable clinicians to administer higher doses of cisplatin without triggering hepatotoxicity.[10]

MAY BE USEFUL FOR CHRONIC RENAL FAILURE

In a study published in 2012 in *International Urology and Nephrology*, researchers from Brazil wanted to determine if fish oil, flaxseed oil, or soybean oil would help male Wistar rats with experimentally induced chronic renal failure. The researchers began by dividing 20 rats with renal failure into four groups of five rats. The rats in the first group served as the controls, the diet of the rats in the second group was supplemented with fish oil, the diet of the rats in the third group was supplemented with flaxseed oil, and the diet of the rats in the fourth group was supplemented with soybean oil. After 30 days, the rats were sacrificed. The researchers learned that both the fish oil and flaxseed oil helped slow the progression of the kidney disease. "Both oils minimized chronic histological injuries in the remnant kidneys of the treated rats."[11]

IS FLAXSEED OIL BENEFICIAL?

In this entry, flaxseed oil appeared to have a number of benefits. People who are dealing with some of the medical problems addressed may wish to discuss flaxseed oil with their medical providers.

NOTES

1. National Center for Complementary and Alternative Medicine. http://nccam.nih.gov.

2. University of Maryland Medical Center. http://www.umm.edu.

3. De Spirt, Silke, Wilhelm Stahl, Hagen Tronnier et al. February 2009. "Intervention with Flaxseed and Borage Oil Supplements Modulates Skin Condition in Women." *British Journal of Nutrition* 101(3): 440–445.

4. Neukam, K., S. De Spirt, W. Stahl et al. 2011. "Supplementation of Flaxseed Oil Diminishes Skin Sensitivity and Improved Skin Barrier Function and Condition." *Skin Pharmacology and Physiology* 24(2): 67–74.

5. de Souza Franco, Eryvelton, Camilla Maria Ferreira de Aquino, Paloma Lys de Medeiros et al. 2012. "Effect of Semisolid Formulation of *Linum usitatissimum* L. (Linseed) Oil

on the Repair of Skin Wounds." *Evidence-Based Complementary and Alternative Medicine* Article ID 270752: 7 pages.

6. Tülüce, Yasin, Halil Özkol, and Ösmail Koyuncu. March 2012. "Photoprotective Effect of Flax Seed Oil (*Linum usitatissimum* L.) against Ultraviolet C-Induced Apoptosis and Oxidative Stress in Rats." *Toxicology and Industrial Health* 28(2): 99–107.

7. Wojtowicz, J. C., I. Butovich, E. Uchiyama et al. March 2011. "Pilot, Prospective, Randomized, Double-Masked, Placebo-Controlled Clinical Trial of an Omega-3 Supplement for Dry Eye." *Cornea* 30(3): 308–314.

8. Salim, Elsayed I., Ahlam E. Abou-Shafey, Ahmed A. Masoud, Salwa A. Elgendy. 2011. "Cancer Chemopreventive Potential of the Egyptian Flaxseed Oil in a Rat Colon Carcinogenesis Bioassay—Implication for Its Mechanism of Action." *Asian Pacific Journal of Cancer Prevention* 12(9): 2385–2392.

9. Tanna, I. R., H. B. Aghera, B. K. Ashok, and H. M. Chandola. January 2012. "Protective Role of *Ashwagandharishta* and Flax Seed Oil against Maximal Electroshock Induced Seizures in Albino Rats." *Pharmacological Research* 33(1): 114–118.

10. Naqshbandi, A., W. Khan, S. Rizwan, and F. Khan. April 2012. "Studies on the Protective Effect of Flaxseed Oil on Cisplatin-Induced Hepatotoxicity." *Human & Experimental Toxicology* 31(4): 364–375.

11. Fernandes, M. B., H. C. Caldas, L. R. Martins et al. October 2012. "Effects of Polyunsaturated Fatty Acids (PUFAs) in the Treatment of Experimental Chronic Renal Failure." *International Urology and Nephrology* 44(5): 1571–1576.

REFERENCES AND RESOURCES

Magazines, Journals, and Newspapers

de Souza Franco, Eryvelton, Camilla Maria Ferreira de Aquino, Paloma Lys de Medeiros et al. 2012. "Effect of a Semisolid Formulation of *Linum usitatissimum* L. (Linseed) Oil on the Repair of Skin Wounds." *Evidence-Based Complementary and Alternative Medicine* Article ID 270752: 7 pages.

De Spirt, Silke, Wilhelm Stahl, Hagen Tronnier et al. February 2009. "Intervention with Flaxseed and Borage Oil Supplements Modulates Skin Condition in Women." *British Journal of Nutrition* 101(3): 440–445.

Fernandes, M. B., H. C. Caldas, L. R. Martins et al. October 2012. "Effects of Polyunsaturated Fatty Acids (PUFAs) in the Treatment of Experimental Chronic Renal Failure." *International Urology and Nephrology* 44(5): 1571–1576.

Naqshbandi, A., W. Khan, S. Rizwan, and F. Khan. April 2012. "Studies on the Protective Effect of Flaxseed Oil on Cisplatin-Induced Hepatotoxicity." *Human & Experimental Toxicology* 31(4): 364–375.

Neukam, K., S. De Spirt, W. Stahl et al. 2011. "Supplementation of Flaxseed Oil Diminishes Skin Sensitivity and Improves Skin Barrier Function and Condition." *Skin Pharmacology and Physiology* 24(2): 67–74.

Salim, Elsayed I., Ahlam E. Abou-Shafey, Ahmed A. Masoud, Salwa A. Elgendy. 2011. "Cancer Chemopreventive Potential of the Egyptian Flaxseed Oil in a Rat Colon Carcinogenesis Bioassay—Implications for Its Mechanism of Action." *Asian Pacific Journal of Cancer Prevention* 12(9): 2385–2392.

Tanna, I. R., H. B. Aghera, B. K. Ashok, and H. M. Chandola. January 2012. "Protective Role of *Ashwagandharishta* and Flax Seed Oil against Maximal Electroshock Induced Seizures in Albino Rats." *Pharmacological Research* 33(1): 114–118.

Tülüce, Yasin, Halil Özkol, and Ösmail Koyuncu. March 2012. "Photoprotective Effect of Flax Seed Oil (*Linum usitatissimum* L.) against Ultraviolet C-Induced Apoptosis and Oxidative Stress in Rats." *Toxicology and Industrial Health* 28(2): 99–107.
Wojtowicz, J.C., I. Butovich, E. Uchiyama et al. March 2011. "Pilot, Prospective, Randomized, Double-Masked, Placebo-Controlled Clinical Trial of an Omega-3 Supplement for Dry Eye." *Cornea* 30(3): 308–314.

Websites

National Center for Complementary and Alternative Medicine. http://nccam.nih.gov.
University of Maryland Medical Center. http://www.umm.edu.

Flaxseed Oil and Cardiovascular Health

As has been noted in the previous entry, some contend that flaxseed oil is useful for a wide variety of medical problems. This entry examines flaxseed oil's potential cardiovascular health benefits.

SUPPORTS CARDIOVASCULAR HEALTH

In a study published in 2012 in the *Journal of Food Science*, researchers from China wanted to determine if flaxseed fortified with vitamin E and phytosterols would support markers of cardiovascular health in rats fed a high-fat diet. The researchers began by dividing their rats into four groups of 10 rats. The diet of the rats in the first group was supplemented with flaxseed oil; the diet of the rats in the second group was supplemented with flaxseed oil fortified with phytosterols. The rats in the third group had a diet supplemented with flaxseed oil and vitamin E; and the rats in the fourth group had a diet supplemented with flaxseed oil, vitamin, and phytosterols. After four weeks, the researchers found that flaxseed oil had lipid-lowering properties. The addition of 2,000mg/100g phytosterols to the flaxseed oil "significantly improved the lipid-lowering activity." Moreover, adding vitamin E to the phytosterols and flaxseed oil "had a synergetic effect on improving plasma oxidative stress and lipid profile in high-fat fed rats."[1]

In a 12-week study published in 2007 in the *European Journal of Clinical Nutrition*, researchers from Athens, Greece, wondered if flaxseed oil would help lower the blood pressure levels of people who had abnormal amounts of lipids in their blood. The cohort initially consisted of 87 men between the ages of 35 and 70. The subjects were randomly assigned to take either 15mL of flaxseed oil per day or 15mL of safflower oil per day. The researchers found that the men taking flaxseed oil experienced improvements in both systolic and diastolic blood pressure readings "by approximately 5mm Hg or 3–6%." They noted that "the magnitude of the hypotensive effect . . . is certainly clinically relevant and is expected to considerably reduce the overall CVD [cardiovascular disease] risk in these patients."[2]

In a study published in 2012 in *Biology and Medicine*, researchers from Morocco noted that coronary heart disease (CHD) has become "a major cause of morbidity and mortality" in postmenopausal women. Because of decreases in the amount of estrogen, "after the onset of menopause, the risk of CHD in women increases dramatically." Would flaxseed oil and/or sesame oil help this medical problem? They tested this notion in female Wistar rats. The researchers randomly placed 32 rats into one of four groups. One group of rats served as the control. To place them in a postmenopausal status, the rats in the three remaining groups had their ovaries removed. The rats in one of these groups were fed a control diet, the rats in the second group were fed 10% flaxseed oil, and the rats in the third group were fed 10% sesame oil. Four weeks later, the rats were sacrificed. After the surgical removal of the ovaries, the researchers found that the rats had increases in their total cholesterol and low-density lipoprotein (LDL or "bad" cholesterol). However, the rats on flaxseed oil and sesame seed oil experienced reductions in both the total cholesterol and LDL. The researchers concluded that flaxseed oil and sesame seed oil "have a beneficial effect on hypercholesterolemia in ovariectomized rats."[3]

In a study published in 2006 in *The Journal of Nutrition*, researchers from Georgia (United States) recruited 56 subjects and randomly placed them on either flaxseed oil or olive oil supplementation. The subjects were evaluated after 12 and 26 weeks. Forty-nine of the subjects completed the study. Most of the subjects were African American women, with an average age of 51. The trial was successfully completed by 27 subjects on flaxseed oil and 22 of the subjects on olive oil. The researchers found that after 12 weeks, the subjects taking flaxseed oil had a 70% increases in plasma alpha-lipoic acid concentrations; no such increase was seen in the subjects on olive oil.[4]

In a study published in 2012 in *Lipids in Health and Disease*, researchers from China wanted to learn if the combination of flaxseed oil and alpha-lipoic acid would reduce the risk for atherosclerosis in rats fed a high-fat diet. The researchers began by dividing 40 male Sprague-Dawley rats into several groups that were fed different combinations of food including both flaxseed oil and alpha-lipoic acid. The researchers found that the combination of flaxseed oil and alpha-lipoic acid supported cardiovascular health. For example, it lowered levels of total cholesterol and LDL, and it was protective against atherosclerosis and reduced inflammation. They concluded that flaxseed oil and alpha-lipoic acid were "effective in amelioration of oxidative stress, lipid profile and inflammation of plasma in rats fed a high-fat diet." And, these two products "might contribute to prevent atherogenesis and then decrease the incidence of CVD."[5]

Meanwhile, a meta-analysis of flaxseed-derived products was conducted by researchers from China, the United Kingdom, and Texas (United States) and published in 2009 in the *American Journal of Clinical Nutrition*. Included in the meta-analysis were 28 studies on flaxseed and flaxseed oil. The researchers found that "whole flaxseed interventions were associated with significant reductions in total and LDL cholesterol, whereas flaxseed oil interventions were not." Treatment with flaxseed oil "induced a modest but non-significant decrease in total and LDL cholesterol compared with baseline values."[6]

The results were similarly modest in a study published in 2011 in the *European Journal of Nutrition*. In this trial, researchers from Colorado (United States) wanted to learn if an eight-week supplementation of a diet with fish oil or flaxseed oil would reduce a certain type of inflammatory factor that increases the risk for stroke and cardiovascular disease. The cohort consisted of 59 healthy adults with an average age of 61. The subjects were randomly placed in one of three groups. Nineteen people were placed in the control group; their diets were supplemented with olive oil. Twenty people had diets supplemented with flaxseed oil, and another 20 had diets supplemented with fish oil. Both treatments failed to trigger any significant reductions in the type of inflammation studied.[7]

In a double-blinded, placebo-controlled clinical trial published in 2008 in the *Journal of the American College of Nutrition*, researchers from Canada tested the cardiovascular benefits of fish oil, flaxseed oil, and hempseed oil in healthy volunteers. More than 80 healthy males and females were randomly assigned to one of four groups. For 12 weeks, they took two 1g capsules per day of sunflower oil (placebo), fish oil, flaxseed oil, or hempseed oil. The researchers learned that none of the oils appeared to support cardiovascular health. "The three oils failed to show any significant effects on lipid profile, LDL oxidation or platelet aggregation." The researchers stressed that "it is important for consumers to know that ingesting two capsules of these oils/day in an attempt to lower lipid levels, reduce LDL oxidation and alter platelet aggregation may not be enough within three months to obtain the desired or expected results."[8]

IS FLAXSEED OIL BENEFICIAL FOR CARDIOVASCULAR HEALTH?

While many people believe that flaxseed oil supports cardiovascular health, the research is not definitive. Before beginning a regime of flaxseed oil for cardiovascular health, it is a good idea to discuss your plans with a medical provider.

NOTES

1. Deng, Qianchun, Xiao Yu, Jiqu Xu et al. June 2012. "Effect of Flaxseed Oil Fortified with Vitamin E and Phytosterols on Antioxidant Defense Capacities and Lipids Profile in Rats." *Journal of Food Science* 77(6): H135–H140.

2. Paschos, G. K., F. Magkos, D. B. Panagiotakos et al. 2007. "Dietary Supplementation with Flaxseed Oil Lowers Blood Pressure in Dyslipidaemic Patients." *European Journal of Clinical Nutrition* 61(10): 1201–1206.

3. Boulbaroud, S., A. El-Hessni, F.-Z. Azzaoui, and A. Mesfioui. 2012. "Sesame Seed Oil and Flaxseed Oil Affect Plasma Lipid Levels and Biomarkers of Bone Metabolism in Ovariectomized Wistar Rats." *Biology and Medicine* 4(3): 102–110.

4. Harper, Charles R., Megan J. Edwards, Andrew P. DeFilipis, and Terry A. Jacobson. January 2006. "Flaxseed Oil Increases the Plasma Concentrations of Cardioprotective (n-3) Fatty Acids in Humans." *The Journal of Nutrition* 136(1): 83–87.

5. Xu, Jiqu, Wei Yang, Qianchun Deng et al. October 31, 2012. "Flaxseed Oil and Alpha-Lipoic Acid Combination Reduces Atherosclerosis Risk Factors in Rats Fed a High-Fat Diet." *Lipids in Health and Disease* 11(1): 148+.

6. Pan, An, Danxia Yu, Wendy Demark-Wahnefried et al. 2009. "Meta-Analysis of the Effects of Flaxseed Interventions on Blood Lipids." *American Journal of Clinical Nutrition* 90(2): 288–297.

7. Nelson, T. L., J. E. Hokanson, and M. S. Hickey. 2011. "Omega-3 Fatty Acids and Lipoprotein Associated Phospholipase A2 in Healthy Older Adult Males and Females." *European Journal of Nutrition* 50(3): 185–193.

8. Kaul, Nalini, Renee Kreml, J. Alejandro et al. 2008. "A Comparison of Fish Oil, Flaxseed Oil and Hempseed Oil Supplementation on Selected Parameters of Cardiovascular Health in Healthy Volunteers." *Journal of the American College of Nutrition* 27(1): 51–58.

REFERENCES AND RESOURCES

Magazines, Journals, and Newspapers

Boulbaroud, S., A. El-Hessni, F.-Z. Azzaoui, and A. Mesfioui. 2012. "Sesame Seed Oil and Flaxseed Oil Affect Plasma Lipid Levels and Biomarkers of Bone Metabolism in Ovariectomized Wistar Rats." *Biology and Medicine* 4(3): 102–110.

Deng, Qianchun, Xiao Yu, Juqu Xu et al. June 2012. "Effect of Flaxseed Oil Fortified with Vitamin E and Phytosterols on Antioxidant Defense Capabilities and Lipids Profile in Rats." *Journal of Food Science* 77(6): H135–H140.

Harper, Charles R., Megan J. Edwards, Andrew P. DeFilipis, and Terry A. Jacobson. 2006. "Flaxseed Oil Increases the Plasma Concentrations of Cardioprotective (n-3) Fatty Acids in Humans." *The Journal of Nutrition* 136(1): 83–87.

Kaul, Nalini, Renee Kreml, J. Alejandro Austria et al. 2008. "A Comparison of Fish Oil, Flaxseed Oil and Hempseed Oil Supplementation on Selected Parameters of Cardiovascular Health in Healthy Volunteers." *Journal of the American College of Nutrition* 27(1): 51–58.

Nelson, T. L., J. E. Hokanson, and M. S. Hickey. 2011. "Omega-3 Fatty Acids and Lipoprotein Associated Phospholipase A2 in Healthy Older Adult Males and Females." *European Journal of Nutrition* 50(3): 185–193.

Pan, An, Danxia Yu, Wendy Demark-Wahnefried et al. 2009. "Meta-Analysis of the Effects of Flaxseed Interventions on Blood Lipids." *American Journal of Clinical Nutrition* 90(2): 288–297.

Paschos, G. K., F. Magkos, D. B. Panagiotakos et al. 2007. "Dietary Supplementation with Flaxseed Oil Lowers Blood Pressure in Dyslipidaemic Patients." *European Journal of Clinical Nutrition* 61(10): 1202–1206.

Xu, Jiqu, Wei Yang, Qianchun Deng et al. October 31, 2012. "Flaxseed Oil and Alpha-Lipoic Acid Combination Reduces Atherosclerosis Risk Factors in Rats Fed a High Fat Diet." *Lipids in Health and Disease* 11(1): 148+.

Websites

National Center for Complementary and Alternative Medicine. http://nccam.nih.gov. University of Maryland Medical Center. www.umm.edu.

Jojoba Oil

Produced from the seeds of the jojoba plant, an evergreen shrub that grows in the desert environments found in southern Arizona and California, northwestern Mexico, the Middle East, and Argentina, jojoba oil (pronounced ho-ho-ba) is 97% liquid wax. Prized for centuries for having a wide variety of medicinal and cosmetic properties, it has been used to heal wounds and open sores. But, it is also found in hair products, moisturizers, and body oils. It is thought to hydrate the scalp and the hair, correct skin dryness, treat brittle nails, and improve skin elasticity. It is believed to have anti-inflammatory, antiseptic, antifungal, and antiaging attributes. In addition, it is often used as a carrier or base oil.[1]

Jojoba oil is readily available in retail stores and online. It is moderately priced.

WOUND-HEALING PROPERTIES

In a study published in 2011 in the *Journal of Ethnopharmacology*, researchers from Italy underscored the fact that wounds are a major medical problem. This is particularly true for wounds that are resistant to healing. "Chronic wounds affect a large number of patients and current worldwide estimates suggest that nearly 6 million people suffer from this kind of disorder." That is why "there is great interest to find new wound healing products." So, the researchers conducted a number of different laboratory studies on human skin cells to assess the wound-healing properties of jojoba liquid wax. The researchers were able to document several of the wound-healing properties of jojoba, and they observed that "it could be proficiently used in the treatment of wounds in clinical settings."[2]

HEALS LESIONS AND ACNE

In an open, prospective, observational pilot study published in 2012 in the Swiss journal *Forschende Komplementärmedizin* (*Research in Complementary Medicine*), researchers from Berlin, Germany, wanted to learn if a jojoba oil facial mask would be useful for people with acne, lesioned skin, or acne-prone skin. For six weeks, 192 females and 2 males applied a facial mask containing clay and jojoba oil two to three times per week. They also completed the required paperwork. As so often happens, everyone did not complete all the study's requirements. Of the initial participants, 133 submitted "complete and precise lesion counts." The researchers found that by the end of the trial, the subjects had a 54% reduction in total lesion count. "Both inflammatory and non-inflammatory skin lesions were reduced significantly after treatment." The researchers concluded that their findings "give preliminary evidence that healing clay jojoba oil facial masks can be effective treatment for lesioned skin and mild acne vulgaris."[3]

EFFECTIVE MOISTURIZER

In a small pilot study published in 2008 in the *Journal of Cosmetic Dermatology*, researchers from Arizona (United States) tested the ability of hydrolyzed jojoba combined with glycerol, which also has hydrating properties, to moisturize human skin. The cohort consisted of nine healthy women, between the ages of 22 and 55. Six of the women were Caucasian; two women were Hispanic; and one woman was African American. Topical treatments of only glycerol or glycerol in combination with different concentrations of hydrolyzed jojoba oil were applied to the lower legs of the women, and evaluations were conducted at baseline and at 8 hours and 24 hours after application. The researchers found that when compared to the areas treated only with glycerol, the areas treated with glycerol and hydrolyzed jojoba oil "resulted in a 40% and 56% decrease in epidermal evaporative water loss values at 8 [hours] and 24h, respectively." The researchers commented that "glycerol and hydrolyzed jojoba esters work in tandem to enhance skin moisturization for at least 24h."[4]

MAY HELP PROTECT AGAINST SCRUB TYPHUS

In a study published in 2010 in *The Southeast Asian Journal of Tropical Medicine and Public Health*, researchers from Malaysia explained that there is a certain type of mite that may bite humans and infect them with scrub typhus. They tested the ability of four different commercial insect repellents to protect against these mites in anesthetized white mice. One of the repellants contained jojoba oil, citronella oil, and tea tree oil. All four of the repellants proved to give protection against the mites. Interesting, "the herbal products were as effective as the product containing DEET."[5]

MAY BE USEFUL TREATMENT FOR CURLY
HAIR THAT IS CHEMICALLY STRAIGHTENED

In a study published in 2008 in the *Journal of Cosmetic Dermatology* researchers from Brazil commented that the chemicals used to straighten very curly hair cause "considerable damage," leaving hair "dry and brittle." As a result, the researchers decided to evaluate the ability of several different products, including jojoba oil, to reduce the amount of harm. The study was conducted on standard Afro-ethnic hair locks. Of the products tested, jojoba oil was one of two products that had the "best results." The researchers noted that incorporating these products "into formulations of straightening products could decrease the damage caused on the threads by the straightening process."[6]

DEMONSTRATES ANTIHERPES
VIRUS PROPERTIES

In a study published in 2010 in *The Open Virology Journal*, researchers from Israel examined the ability of jojoba ethanol and aquatic extracts and another plant-based product (inch plant) to destroy various types of herpes viruses in a laboratory setting. Why is that important? Because acyclovir and similar anti-herpes medications have unpleasant side effects such as nausea, vomiting, head-aches, rash, and diarrhea. In addition, there have emerged strains of the herpes virus that are resistant to medications. Moreover, these medications are not nec-essarily effective in viral attacks that reoccur.

When the researchers applied various concentrations of the jojoba treat-ments at different times, they found that jojoba extract inhibited the growth of herpes viruses. "The highest antiviral activity of the extracts against all tested viruses was obtained when the cells were treated with the extract at the time [of infection] and post infection." And, the researchers concluded that "the extracts tested in this study may therefore provide a potential source of effective anti-herpetic compounds which seem less toxic than ACV [acyclovir]."[7]

ANTI-INFLAMMATION PROPERTIES

In a study published in 2005 in *Pharmacological Research*, researchers from Egypt tested the potential anti-inflammatory activity of jojoba liquid wax in a number of experimental models that used adult male Sprague-Dawley rats. Among these tests were measurements of the anti-inflammation properties of jojoba in the paws and the ears of rats with experimentally induced inflamed paws and ears. The researchers found that jojoba liquid wax had notable anti-inflammatory "activity in several animal models." They noted that their "results lend support to the effectiveness of JLW [jojoba liquid wax] in combating inflam-mation via multilevel regulation of inflammatory mediators."[8]

POTENTIAL ALLERGIC REACTION
TO JOJOBA OIL

Allergic reactions to jojoba oil are believed to be quite rare. Nevertheless, such responses may occur. For example, in a brief article published in 2006 in *Contact Dermatitis*, researchers from Italy described a "delayed hypersensitiv-ity" to jojoba in a 43-year-old female patient with a history of allergies. The woman reported that one day earlier, she had applied cosmetic body cream to her skin. Within 24 hours, she had red blistering lesions spread over her body. After about seven days of treatment with an oral corticosteroid, she recovered. The researchers used patch tests to determine the woman's sensitivity to ingredients in the body cream. Only the test for jojoba was positive. When the patch test was repeated four months later, the same positive response was observed.[9]

IS JOJOBA OIL BENEFICIAL?

Although not as well studied as other oils, jojoba oil appears to have a number of favorable properties. It seems to be particularly useful for some skin-related problems.

NOTES

1. Shaath, Nadim A. September 2012. "The Wonders of Jojoba: This Natural Material Has a Wide Range of Application in Personal Care Products." *Household & Personal Products Industry* 49(9): 47–50.

2. Ranzato, Elia, Simona Martinotti, and Bruno Burlando. March 2011. "Wound Healing Properties of Jojoba Liquid Wax: An *In Vitro* Study." *Journal of Ethnopharmacology* 134(2): 443–449.

3. Meier, L., R. Stange, A. Michalsen, and B. Uehleke. 2012. "Clay Jojoba Oil Facial Mask for Lesioned Skin and Mild Acne—Results of a Prospective, Observational Pilot Study." *Forschende Komplementärmedizin* (*Research in Complementary Medicine*) 19(2): 75–79.

4. Meyer, J., B. Marshall, M. Gacula Jr., and L. Rheins. December 2008. "Evaluation of Additive Effects of Hydrolyzed Jojoba (*Simmondsia chinensis*) Esters and Glycerol: A Preliminary Study." *Journal of Cosmetic Dermatology* 7(4): 268–274.

5. Hanifah, A. L., S. H. Ismail, and H. T. Ming. September 2010. "Laboratory Evaluation of Four Commercial Repellents against Larval *Leptotrombidium deliense* (Acari: Trombiculidae)." *The Southeast Asian Journal of Tropical Medicine and Public Health* 41(5): 1082–1087.

6. Dias, T.C., A.R. Baby, T.M. Kaneko, and M.V. Velasco. June 2008. "Protective Effect of Conditioning Agents on Afro-Ethnic Hair Chemically Treated with Thioglycolate-Based Straightening Emulsion." *Journal of Cosmetic Dermatology* 7(2): 120–126.

7. Yarmolinsky, Ludmila, Michele Zaccai, Shimon Ben-Shabat, and Mahmoud Huleihel. 2010. "Anti-Herpetic Activity of *Callissia fragrans* and *Simmondsia chinensis* Leaf Extracts *In Vitro*." *The Open Virology Journal* 4(1): 57–62.

8. Habashy, Ramy R., Ashraf B. Abdel-Naim, Amani E. Khalifa, and Mohammed M. Al-Azizi. 2005. "Anti-Inflammatory Effects of Jojoba Liquid Wax in Experimental Models." *Pharmacological Research* 51(2): 95–105.

9. Di Berardino, L., F. Di Berardino, A. Castelli, and F. Della Torre. 2006. "A Case of Contact Dermatitis from Jojoba." *Contact Dermatitis* 55(1): 57, 58.

REFERENCES AND RESOURCES
Magazines, Journals, and Newspapers

Baumann, Leslie S. October 2010. "Jojoba." *Skin & Allergy News* 41(10): 18.

Di Berardino, L., F. Di Berardino, A. Castelli, and F. Della Torre. 2006. "A Case of Contact Dermatitis from Jojoba." *Contact Dermatitis* 55(1): 57, 58.

Dias, T. C., A. R. Baby, T. M. Kaneko, and M. V. Velasco. June 2008. "Protective Effect of Conditioning Agents on Afro-Ethnic Hair Chemically Treated with Thioglycolate-Based Straightening Emulsion." *Journal of Cosmetic Dermatology* 7(2): 120–126.

Habashy, Ramy R., Ashraf B. Abdel-Naim, Amani E. Khalifa, and Mohammed M. Al-Azizi. 2005. "Anti-Inflammatory Effects of Jojoba Liquid Wax in Experimental Models." *Pharmacological Research* 51(2): 95–105.

Hanifah, A. L., S. H. Ismail, and H. T. Ming. September 2010. "Laboratory Evaluation of Four Commercial Repellants against Larval *Leptotrombidium deliense* (Acari: Trombiculidae)." *The Southeast Asian Journal of Tropical Medicine and Public Health* 41(5): 1082–1087.

Meier, L., R. Stange, A. Michalsen, and B. Uehleke. 2012. "Clay Jojoba Oil Facial Mask for Lesioned Skin and Mild Acne—Results of a Prospective, Observational Pilot Study." *Forschende Komplementärmedizin* (*Research in Complementary Medicine*) 19(2): 75–79.

Meyer, J., B. Marshall, M. Gacula Jr., and L. Rheins. December 2008. "Evaluation of Additive Effects of Hydrolyzed Jojoba (*Simmondsia chinensis*) Esters and Glycerol: A Preliminary Study." *Journal of Cosmetic Dermatology* 7(4): 268–274.

Ranzato, Elia, Simona Martinotti, and Bruno Burlando. March 24, 2011. "Wound Healing Properties of Jojoba Liquid Wax: An *In Vitro* Study." *Journal of Ethnopharmacology* 134(2): 443–449.

Shaath, Nadim A. September 2012. "The Wonders of Jojoba: This Natural Material Has a Wide Range of Application in Personal Care Products." *Household & Personal Products Industry* 49(9): 47–50.

Yarmolinsky, Ludmila, Michele Zaccai, Shimon Ben-Shabat, and Mahmoud Huleihel. 2010. "Anti-Herpetic Activity of *Callissia fragrans* and *Simmondsia chinensis* Leaf Extracts *In Vitro*." *The Open Virology Journal* 4(1): 57–62.

Website

WebMD. www.webmd.com.

Krill Oil

Until a few decades ago, very few people had ever heard of krill oil. Even fewer knew anything about it. And, krill oil was difficult to locate. People who wished to consume krill oil had very few choices. More recently, it has been featured in many articles and television programs. It is now readily available in stores and online. But, it is somewhat pricey.

Krill oil is made from small, red-colored crustaceans that live in extremely cold ocean water, such as the Arctic Ocean. They contain high amounts of long-chain polyunsaturated fatty acids, such as eicosapentaenoic acid and docosahexaenoic acid, thereby making them a bountiful source of omega-3 fatty acids. Krill oil has high amounts of the potent antioxidant called astaxanthin and small amounts of vitamins A, D, and E.

Krill oil is said to support cardiovascular health by reducing cholesterol and levels of inflammation. It is also thought to lower the incidence of clots and the buildup of plaque in the vascular walls. Additionally, it may be useful for arthritis and other musculoskeletal disorders.[1]

CARDIOVASCULAR HEALTH

In a 12-week double-blind, randomized study published in 2004 in *Alternative Medicine Review*, researchers from Canada wanted to assess the effect krill oil had on blood lipids—total cholesterol, triglycerides, low-density lipoprotein (LDL or "bad" cholesterol), and high-density lipoprotein (HDL or "good" cholesterol). The cohort consisted of 120 people between the ages of 18 and 75 who had elevated levels of cholesterol and triglycerides for at least six months. The participants were assigned to one of four groups. Thee members of groups A and B took different doses of krill supplementation, the members of group C took fish oil, and the members of group D took a placebo. By the end of the study, the participants taking krill oil exhibited the most improvement in their lipids. The researchers noted that their findings "demonstrate within high levels of confidence that krill oil is effective for the management of hyperlipidemia [high lipid levels] by significantly reducing total cholesterol, LDL and triglycerides and increasing HDL levels."[2]

In a study published in 2012 in the *Journal of Animal Physiology and Animal Nutrition*, researchers from Italy divided rats into three groups. The control rats were fed only rat food. A second group of rats received regular rat food plus krill oil, and a third group of rats ate rat food with fish oil. After six weeks of supplementation, the rats taking krill oil had a 33% reduction in cholesterol levels; the rats taking the fish oil had a 21% reduction. Reductions in triglycerides were 20% for krill oil and 10% for fish oil.[3]

On the other hand, different results were obtained in a study published in 2011 in *Lipids*. Researchers from Norway and Sweden examined the effects of krill oil and fish oil on serum lipids and markers of oxidative stress and inflammation. The cohort consisted of 113 subjects who had normal or slightly elevated total blood cholesterol and/or triglycerides levels. They were randomized into one of three groups—seven weeks of krill oil, fish oil, or no supplementation (control). One hundred and fifteen people completed the study. The rates of withdrawal were similar in all three groups. Yet, no differences were observed. Similar results were obtained from both the krill and fish oils—though the dose of krill oil was 62.8% of that of the fish oil. So, the same results were obtained with less krill oil. The researchers commented that "no statistically significant differences in changes in any of the serum lipids or the markers of oxidative stress and inflammation between the study groups were observed."[4]

EFFECTIVE FOR CHRONIC INFLAMMATION AND ARTHRITIC SYMPTOMS

In a randomized, double-blind, placebo-controlled study published in 2007 in the *Journal of the American College of Nutrition*, a researcher from Canada wanted to determine if krill oil was useful for chronic inflammation and arthritic symptoms. The cohort consisted of 90 people between the ages of 30 and 75. All the participants had confirmed diagnoses of cardiovascular disease and/or rheumatoid arthritis and/or osteoarthritis, and they all had elevated levels of C-reactive protein, a protein produced by the liver that rises when there is inflammation in the body. The participants were assigned to one of two groups. For 30 days, the members of one group took 300mg of krill oil each morning; the members of the other took placebos. For the millions suffering from chronic inflammation and arthritic symptoms, the results are certainly notable. According to the researcher, "a daily dose of 300mg significantly inhibits inflammation and reduces arthritic symptoms within a short treatment periods of 7 and 14 days."[5]

USEFUL FOR RHEUMATOID ARTHRITIS

In a 68-day study published in 2010 in *BMC Musculoskeletal Disorders*, researchers from Switzerland, Norway, and Finland investigated the use of krill and fish oils in mice that were prone to develop rheumatoid arthritis. The cohort initially consisted of three groups; each had 14 mice. One group ate supplemental krill oil, second group had supplemental fish oil; and the third group served as the control. But, not all the mice were able to complete the study. Still, the researchers found that when compared to the mice on the control diet, the mice taking krill and fish oils had reductions in their symptoms of arthritis. However, by the last phase of the study, only the krill oil provided significantly levels of relief. The researchers noted that "krill oil was able to reduce the severity of arthritis by about 50%." They concluded that "krill oil provides protection in terms of arthritis scores and joint pathology in the CIA [collagen-induced arthritis] model." And, they added that krill oil may prove to be useful for rheumatoid arthritis as well as other types of arthritis and other inflammatory conditions.[6]

MAY HELP PEOPLE WITH ULCERATIVE COLITIS

In a four-week study published in 2012 in the *Scandinavian Journal of Gastroenterology*, researchers wanted to learn if krill oil would be useful for ulcerative colitis, a type of inflammatory bowel disease that affects the lining of the large intestine and rectum. The researchers, who were based in Norway, began by dividing 30 rats into three equal groups. One group was fed a standard rat diet; a second group, which had dextran sulfate sodium–induced colitis in the final week, was also fed a standard diet. The rats in the third group had a similarly induced colitis in the final week, but their diet included standard food plus 5% krill oil. When they compared the two groups that had induced colitis, the researchers found that the rats in the krill oil group demonstrated preservation of the colon length, a reduction in markers of oxidative stress, and a number of different beneficial changes. They noted that their findings "indicate an anti-inflammatory and protein antioxidant effect of KO [krill oil]."[7]

MAY BE USEFUL FOR MANAGEMENT OF PREMENSTRUAL SYNDROME AND DYSMENORRHEA

In a study published in 2003 in *Alternative Medicine Review*, researchers from Quebec, Canada, evaluated the ability of Neptune Krill Oil to help manage the symptoms of premenstrual syndrome and dysmenorrhea (pain from menstruation), and they compared the effectiveness of Neptune Krill Oil to omega-3 fish oil. The cohort consisted of 70 women with a mean age of 33. For three months, the women took either Neptune Krill Oil or omega-3 fish oil supplementation. The results were notable. The researchers wrote that the krill oil "can significantly reduce dysmenorrhea and the emotional symptoms of premenstrual syndrome and is shown to be significantly more effective for the complete management of premenstrual symptoms compared to omega-3 fish oil."[8]

IS KRILL OIL BENEFICIAL?

Krill oil appears to have a number of potential benefits. Future research should yield even more definitive information.

NOTES

1. Beer, Christina. October 2011. "Krill: The Ocean's Gold: Krill Is One of the Ocean's Greatest Treasures and Consumers Are Just Beginning to Appreciate Its Value Too." *Nutraceuticals World* 14(8): 62–64.

2. Bunea, Ruxandra, Khassan El Farrah, and Luisa Deutsch. December 2004. "Evaluation of the Effects of Neptune Krill Oil on the Clinical Course of Hyperlipidemia." *Alternative Medicine Review* 9(4): 420–428.

3. Ferramosca, A., L. Conte, and V. Zara. April 2012. "A Krill Oil Supplemented Diet Reduces the Activities of the Mitochondrial Tricarboxylate Carrier and of the Cytosolic Lipogenic Enzymes in Rats." *Journal of Animal Physiology and Animal Nutrition* 96(2): 295–306.

4. Ulven, Stine, Bente Kirkhus, Amandine Lamglait et al. 2011. "Metabolic Effects of Krill Oil Are Essentially Similar to Those of Fish Oil but at Lower Dose of EPA and DHA, in Healthy Volunteers." *Lipids* 46(1): 37–46.

5. Deutsch, Luisa. February 2007. "Evaluation of the Effect of Neptune Krill Oil on Chronic Inflammation and Arthritic Symptoms." *Journal of the American College of Nutrition* 26(1): 39–48.

6. Ierna, Michelle, Alison Kerr, Hannah Scales et al. June 29, 2010. "Supplementation of Diet with Krill Oil Protects against Experimental Rheumatoid Arthritis." *BMC Musculoskeletal Disorders* 11(1): 136–146.

7. Grimstad, Tore, Bodil Bjørndal, Daniel Cacabelos et al. January 2012. "Dietary Supplementation of Krill Oil Attenuates Inflammation and Oxidative Stress in Experimental Ulcerative Colitis in Rats." *Scandinavian Journal of Gastroenterology* 47(1): 49–58.

8. Sampalis, F., R. Bunea, M. F. Pelland et al. May 2003. "Evaluation of the Effects of Neptune Krill Oil on the Management of Premenstrual Syndrome and Dysmenorrhea." *Alternative Medicine Review* 8(2): 171–179.

REFERENCES AND RESOURCES
Journals, Magazines, and Newspapers

Beer, Christina. October 2011. "Krill: The Ocean's Gold: Krill Is One of the Ocean's Greatest Treasures and Consumers Are Just Beginning to Appreciate Its Value Too." *Nutraceuticals World* 14(8): 62–64.

Bunea, Ruxandra, Khassan El Farrah, and Luisa Deutsch. December 2004. "Evaluation of the Effects of Neptune Krill Oil on the Clinical Course of Hyperlipidemia." *Alternative Medicine Review* 9(4): 420–428.

Deutsch, Luisa. February 2007. "Evaluation of the Effect of Neptune Krill Oil on Chronic Inflammation and Arthritis Symptoms." *Journal of the American College of Nutrition* 26(1): 39–48.

Ferramosca, A., L. Conte, and V. Zara. April 2012. "A Krill Supplemented Diet Reduces the Activities of the Mitochondrial Tricarboxylate Carrier and of the Cytosolic Lipogenic Enzymes in Rats." *Journal of Animal Physiology and Animal Nutrition* 96(2): 295–306.

Grimstad, Tore, Bodil Bjørndal, Daniel Cacabelos et al. January 2012. "Dietary Supplementation of Krill Oil Attenuates Inflammation and Oxidative Stress in Experimental Ulcerative Colitis in Rats." *Scandinavian Journal of Gastroenterology* 47(1): 49–58.

Ierna, Michelle, Alison Kerr, Hannah Scales et al. June 29, 2010. "Supplementation of Diet with Krill Oil Protects against Experimental Rheumatoid Arthritis." *BMC Musculoskeletal Disorders* 11(1): 136–146.

Sampalis, F., R. Bunea, M. E. Pelland et al. May 2003. "Evaluation of the Effects of Neptune Krill Oil on the Management of Premenstrual Syndrome and Dysmenorrhea." *Alternative Medicine Review* 8(2): 171–179.

Ulven, Stine, Bente Kirkhus, Amandine Lamglait et al. 2011. "Metabolic Effects of Krill Oil Are Essentially Similar to Those of Fish Oil but at Lower Dose of EPA and DHA, in Healthy Volunteers." *Lipids* 46(1): 37–46.

Website

University of Maryland Medical Center. www.umm.edu.

Olive Oil: Overview

One of the oldest foods, olives are thought to have originated in Crete or Syria between 5,000 and 7,000 years ago. Olive oil, which is made from the crushing and pressing of olives, has been consumed since at least 3,000 BC. The Spanish colonizers brought olive tress to North America between the 1500s and 1700s.[1]

Although all varieties of olive oil are often lumped together, there are actually a number of different types. Extra-virgin olive oil is the highest quality and most flavorful type of olive oil. It is produced without the use of any additives or solvents and in temperatures that do not harm the oil in any way. When purchasing olive oil, it is always best to buy extra-virgin olive oil. Other varieties of olive oil include virgin olive oil, ordinary virgin olive oil, olive oil, and refined olive oil.

Olive oil needs to be protected from light. That is why it is best to buy oil in tinted glass bottles. Once a bottle of olive oil is opened, it should be used within a two-month period of time. In addition, since heat may degrade the contents, try not to store olive oil next to the stove.

Olive oil is about 72% oleic acid, a monounsaturated fatty acid. And, it contains three categories of phenols, which inhibit oxidation.[2]

Olive oil is believed to be useful for a wide variety of medical concerns. It is said to support cardiovascular health and prevent the growth of cancer cells. In addition, it is thought to be useful for people with diabetes and for those who are obese. And, it is also considered to be helpful for people with arthritic conditions such as rheumatoid arthritis and for those dealing with bone loss from osteopenia and osteoporosis. Moreover, it may provide protection against viruses and has been used in skin care for thousands of years.

Widely available in traditional and specialty stores, olive oil is also sold in many online stores. Prices vary, but a good quality extra-virgin olive oil may be purchased at a moderate price. For most people, it is not necessary to buy very expensive olive oil. But, do avoid the truly inexpensive olive oil. Inevitably, you will be disappointed with the taste. When buying olive oil, look for the "Harvest Date" on the label. If no harvest date is listed, select another brand. You will want to use the entire bottle of olive oil within two years of the harvest date.

This entry examines a variety of olive oil's purported benefits, while subsequent entries explore specific conditions that olive oil may help with in more detail.

MAY BE USEFUL FOR NEUROLOGICAL ILLNESSES SUCH AS HUNTINGTON'S DISEASE

In a study published in 2011 in *Nutritional Neuroscience*, researchers from Spain wanted to learn if extra-virgin olive oil and hydroxytyrosol, a polyphenol found in olive oil, could help reduce the oxidative brain damage that occurs in neuro-degenerative diseases like Huntington's disease. They divided their 40 Wistar rats into different groups, and rats were treated with various combinations Hunting-ton's disease–inducing chemicals, extra-virgin olive oil, and hydroxytyrosol. One group of rats was set aside to serve as the control. The researchers found that both extra-virgin olive oil and hydroxytyrosol "exerted strong antioxidative effects" against the chemical causing the symptoms of Huntington's disease.[3]

MAY HELP HEAL SPINAL INJURIES

In a study published in 2012 in *Neurological Research*, researchers from Iran investigated the ability of oleuropein, a polyphenol found in olive oil, to help heal experimental spinal cord injuries in rats. After dividing their rats into sev-eral groups, the researchers injured the spines of most of the rats. Some of the rats were treated with oleuropein. The researchers learned that the oleuropein pro-vided a degree of protection from spinal cord injury. "All the findings obtained from the present study demonstrated biochemical, histopathological, and func-tional evidences that OE [oleuropein] treatment after SCI [spinal cord injury] had neuroprotective effects."[4]

MAY REDUCE WHEEZING DURING BABY'S FIRST YEAR

In a study published in 2010 in *Pediatric Pulmonology*, researchers from Spain and Chile wanted to learn if a pregnant mother's consumption of olive oil in cooking and dressings for salads and vegetables would reduce the amount of wheezing experienced by her newborn during his or her first year of life. The co-hort included 1,409 infants who attended healthy infant clinics in Spain. The re-searchers learned that when mothers-to-be used olive oil as their primary source of cooking and salad dressing oil, their newborns had almost a 40% reduction in wheezing during the first year. The mothers in the wheezing group used more margarine, butter, and other oils. The researchers noted that their findings "could suggest that the increase of olive oil consumption during pregnancy might be a primary protective measure to reduce wheezing early in life."[5]

Meanwhile, researchers from Greece and Spain reviewed five studies on the as-sociation between the consumption of a Mediterranean diet by pregnant women and children and the incidence of asthma and allergies in children. Four of these studies were conducted in Greece and Spain; one was conducted in Mexico. The results appeared in a 2009 article in *Public Health Nutrition*. The research-ers found that "all of the studies reported beneficial associations between a high level of adherence to the Mediterranean diet during childhood and symptoms of asthma or allergic rhinitis." Two of the studies focused on eating a prenatal Mediterranean diet. One of these studies included a cohort of 460 children from Spain; these children were analyzed during a follow-up that was held when they were 6.5 years old. This study found that a high adherence to a Mediterranean diet during pregnancy was protective against wheezing and allergies. The second study, which was conducted in Mexico, had a cross-sectional format and included 1,476 children between the ages of six and seven years. This study found no association between the mother's adherence to a Mediterranean diet and wheezing and allergies in the children. However, the researchers noted that the "associations should be interpreted cautiously because of the reliability of maternal recall of diet 6–7 years previously."[6]

MAY BE USEFUL FOR DEPRESSION

In a study published in 2011 in *PLoS One*, researchers from Spain and the Netherlands evaluated the intake of fatty acids and culinary fats and the incidence of depression in a Mediterranean population. The initial cohort consisted of 12,059 Spanish university graduates with a mean age of 37.5 years who were free of depression. During the follow-up of about six years, a total of 657 new cases of depression were identified. The researchers found that the intake of olive oil was associated with a lower incidence of depression. Thus, olive oil "consumption was inversely associated with depression risk."[7]

In a prospective epidemiological investigation published a few years earlier, in 2009, in the *Journal of Psychiatric Research*, researchers from Greece and Massachusetts (United States) wanted to "identify dietary lipids potentially associated with affective state and depression." The initial cohort consisted of 610 healthy men and women aged 60 years or older, who lived in Greece. Between 6 and 13 years after the study began, the participants were evaluated for depression. The researchers found that olive oil was associated with a decreased risk for depression; at the same time, they determined that seed oil was associated with an increased risk for depression. They concluded that a "lower intake of seed oils and higher intake of olive oil prospectively predict a healthier affective state."[8]

In a study published in 2013 in *ISRN Pharmacology*, researchers from Pakistan examined the ability of extra-virgin olive oil to reduce depression and anxiety in male albino Wistar rats. The rats in the test group were treated with olive oil every day for four weeks; the control rats were given an equal amount of water. The researchers had the rats swim, which increased depressive symptoms, and placed them in an apparatus that made them afraid of the "novel and open areas," a situation that naturally increased their anxiety. Rats that had been treated with the extra-virgin olive oil had fewer depressive symptoms and less anxiety. The researchers concluded that extra-virgin olive oil may "be used as a therapeutic substance for the treatment of depression and anxiety."[9]

MAY BE USEFUL FOR ARTHRITIS

In an eight-week trial published in 2013 in *The Journal of Nutritional Biochemistry*, researchers from Italy and Austria wanted to learn if olive oil would be useful for people dealing with osteoarthritis, a very common degenerative joint disease characterized by damage to the cartilage, the material that cushions the joints. In osteoarthritis, joints swell and are chronically inflamed. When it occurs in the hand joints, people find it harder to move their hands; when it occurs in the knee joints, people find it harder to walk and exercise.

The researchers hypothesized that "the supplementation of extra-virgin olive oil could modify the natural history of joint injury, reducing inflammation and promoting lubrication." Moreover, when combined with a "mild" amount of physical activity, "diet could help to improve chromic inflammation and osteo-arthritis occurrence." The researchers divided their 45 rats into five groups. The rats in the first group served as the control, the rats in the second group had injured cartilage, the rats in the third group had injured cartilage and were ex-ercised on a treadmill, the rats in the fourth group had injured cartilage and a diet supplemented with extra-virgin olive oil, and the rats in the fifth group had injured cartilage, treadmill training, and a diet supplemented with extra-virgin olive oil. The researchers' hypotheses proved valid; extra-virgin olive oil repeat-edly demonstrated strong anti-inflammatory properties in the rats. The research-ers commented that their findings "confirmed the importance of physical activity (mild exercise) in conjunction with extra-olive oil diet in medical therapy to prevent OA [osteoarthritis] disease."[10]

IS OLIVE OIL BENEFICIAL?

From the research reviewed, it appears that olive oil, especially extra-virgin olive oil, has a host of healthful properties. Unless people have an allergy or in-tolerance to olives or olive oil, conditions that are thought to be exceedingly rare, extra-virgin olive oil should be included in the diet. In fact, it should serve the role of the most important oil in the diet.

NOTES

1. The George Mateljan Foundation. www.whfoods.com.

2. Lockwood, Brian and Emily Waterman. December 2007. "Active Components and Clinical Application of Olive Oil." *Alternative Medicine Review* 12(4): 331–342.

3. Tasset, I., A.J. Pontes, A.J. Hinojosa et al. 2011. "Olive Oil Reduces Oxidative Damage in a 3-Nitropropionic Acid-Induced Huntington's Disease-Like Rat Model." *Nutritional Neuroscience* 14(3): 106–111.

4. Khalatbary, Ali Reza and Hassan Ahmadvand. 2012. "Neuroprotective Effects of Oleuropein Following Spinal Cord Injury in Rats." *Neurological Research* 34(1): 44–51.

5. Castro-Rodriguez, Jose A., Luis Garcia-Marcos, Manuel Sanchez-Solis et al. 2010. "Olive Oil during Pregnancy Is Associated with Reduced Wheezing during the First Year of Life of the Offspring." *Pediatric Pulmonology* 45(4): 395–402.

6. Chatzi, Leda and Manolis Kogevinas. 2009. "Prenatal and Childhood Mediterra-nean Diet and the Development of Asthma and Allergies in Children." *Public Health Nutrition* 12(9A): 1629–1634.

7. Sánchez-Villegas, Almudena, Lisa Verberne, Jokin De Irala et al. January 2011. "Dietary Fat Intake and the Risk of Depression: The SUN Project." *PLoS ONE* 6(1): e16268.

8. Kyrozis, A., T. Psaltopoulou, P. Stathopoulos et al. May 2009. "Dietary Lipids and Geriatric Depression Scale Score among Elders: The EPIC–Greece Cohort." *Journal of Psychiatric Research* 43(8): 763–769.

9. Perveen, T., B.M. Hashmi, S. Haider et al. July 10, 2013. "Role of Monoaminergic System in the Etiology of Olive Oil Induced Antidepressant and Anxiolytic Effects in Rats." *ISRN Pharmacology* Article ID 615685: 5 pages.

10. Musumeci, G., F.M. Trovato, K. Pichler et al. December 2013. "Extra-Virgin Olive Oil Diet and Mild Physical Activity Prevent Cartilage Degeneration in an Osteoarthritis Model: An *In Vivo* and *In Vitro* Study on Lubricin Expression." *The Journal of Nutritional Biochemistry* 24(12): 2064–2075.

REFERENCES AND RESOURCES

Magazines, Journals, and Newspapers

Castro-Rodriguez, Jose A., Luis Garcia-Marcos, Manuel Sanchez-Solis et al. April 2010. "Olive Oil during Pregnancy Is Associated with Reduced Wheezing during the First Year of Life of the Offspring." *Pediatric Pulmonology* 45(4): 395–402.

Chatzi, Leda and Manolis Kogevinas. 2009. "Prenatal and Childhood Mediterranean Diet and the Development of Asthma and Allergies in Children." *Public Health Nutrition* 12(9A): 1629–1634.

Khalatbary, Ali Reza and Hassan Ahmadvand. 2012. "Neuroprotective Effect of Oleuropein Following Spinal Cord Injury in Rats." *Neurological Research* 34(1): 44–51.

Kyrozis, A., T. Psaltopoulou, P. Stathopoulos et al. 2009. "Dietary Lipids and Geriatric Depression Scale Score among Elders: The EPIC-Greece Cohort." *Journal of Psychiatric Research* 43: 763–769.

Lockwood, Brian and Emily Waterman. December 2007. "Active Components and Clinical Application of Olive Oil." *Alternative Medicine Review* 12(4): 331–342.

Musumeci, G., F.M. Trovato, K. Pichler et al. December 2013. "Extra-Virgin Olive Oil Diet and Mild Physical Activity Prevent Cartilage Degeneration in an Osteoarthritis Model: An *In Vivo* and *In Vitro* Study on Lubricin Expression." *The Journal of Nutritional Biochemistry* 24(12): 2064–2075.

Perveen, T., B.M. Hashmi, S. Haider et al. July 10, 2013. "Role of Monoaminergic System in the Etiology of Olive Oil Induced Antidepressant and Anxiolytic Effects in Rats." *ISRN Pharmacology* Article ID 615685: 5 pages.

Sánchez-Villegas, Almudena, Lisa Verberne, Jokin De Irala et al. January 2011. "Dietary Fat Intake and the Risk of Depression: The SUN Project." *PLoS ONE* 6(1): e16268.

Tasset, I., A.J. Pontes, A.J. Hinojosa et al. 2011. "Olive Oil Reduces Oxidative Damage in a 3-Nitropropionic Acid-Induced Huntington's Disease-Like Rat Model." *Nutritional Neuroscience* 14(3): 106–111.

Websites

The George Mateljan Foundation. www.whfoods.com.

Olive Oil Times. www.oliveoiltimes.com.

Olive Oil and Bone Health

This entry examines studies related to olive oil's benefits for bone health.

PROTECTS BONES

In a study published in 2009 in *Nutrition*, researchers from Athens, Greece, wanted to learn if adherence to a Mediterranean diet, including olive oil, or other dietary patterns had any significant effect on bone health. The cohort consisted of 220 adult Greek women, with a mean age of 48. Researchers reviewed their diets and conducted tests to evaluate their bone health. The researchers found that the women who followed a diet with features of the Mediterranean diet, such as "rich in fish and olive oil," had improved bone health. They concluded that a "high consumption of fish and olive oil and low red meat intake was positively related to bone mass, suggesting potential bone-preserving properties of this pattern throughout adult life."[1]

In a two-part crossover study published in 2008 in the *Mediterranean Journal of Nutrition and Metabolism*, researchers from Italy examined the bone status of 15 women, between the ages of 25 and 40, following two dietary phases. During the first three-week phase, the subjects took 20mL/day of organic standard extra-virgin olive oil. During the second three-week phase, the subjects added 20mL/day of an extra-virgin olive oil that contained vitamin K to their diet. The researchers found that the extra-virgin oil helped prevent bone loss in the volunteers, and they concluded that the olive oil they used "might be useful for bone protection."[2]

In a study published in *BMC Complementary & Alternative Medicine*, researchers from Egypt divided adult female Wistar rats between the ages of 12 and 14 months into three groups. To induce menopause and increase the risk of bone loss, the rats in two of the groups had surgeries to remove their ovaries. For four weeks before this surgery, known as an ovariectomy, and eight weeks after the procedure, one group of ovariectomized rats received extra-virgin olive oil supplementation. The researchers found that the olive oil prevented some of the markers for bone loss experienced by the ovariectomized rats. For example, it raised levels of calcium in the blood and "attenuated ovariectomy-induced osteoporosis." In addition, olive oil "prevented bone loss and decreased resorption of bone." According to the researchers, "olive oil represents a promising therapeutic option for the prevention and/or treatment of postmenopausal osteoporosis."[3]

In a study published in 2012 in *The Journal of Clinical Endocrinology and Metabolism*, researchers from Spain evaluated the association between circulating bone formation and resorption markers and the intake of olive oil. The cohort consisted of 127 community-dwelling men between the ages of 55 and 80. They were randomly placed on a low-fat diet or a Mediterranean diet with three daily tablespoons of olive oil or a Mediterranean diet with about one ounce of mixed nuts per day. The participants completed questionnaires about their lifestyles, medical conditions, and use of medications at baseline, after one year, and again after two years. At the end of the study, the researchers learned that the men who ate the Mediterranean diet enriched with olive oil displayed higher levels of osteocalcin in their blood, a known marker for strong and healthy bones. No such elevation of osteocalcin levels occurred in the other two groups of men. "The main finding is that consumption of a MedDiet enriched with olive oil, but not a MedDiet enriched with nuts or a control [low-fat diet] was associated with a significant increase of total osteocalcin concentrations." Thus, the researchers concluded that their findings demonstrated that olive oil protects bones.[4]

In a study published in 2006 in *Clinical Nutrition*, researcher from France, Greece and the Ivory Coast examined the ability of oleuropein, a major polyphenol in olive leaves and olive oil, to support the bone health of rats. The researchers began by dividing their 98 rats into two groups. Twenty rats were set aside to serve as controls; 78 rats had the surgical removal of their ovaries. After surgery, these rats were divided into several groups. The control rats and some of the ovariectomized rats were fed a standard diet. But, 52 of the ovariectomized rats were divided into four groups that were fed varying amounts of supplemental oleuropein. The researchers found that all the doses of oleuropein reduced the amount of ovariectomy-induced bone loss. "Data obtained in the present study (carried out in 13 animals per group) confirm the bone sparing effect of oleuropein in the experimental model for senile osteoporosis."[5]

Most of the researchers in the previous study joined with a researcher from Greece to conduct a similar study that was published in 2008 in the *Journal of Agricultural and Food Chemistry*. In this study, the researchers wanted to learn if the consumption of 84 days of tyrosol and hydroxytyrosol, "the main olive oil phenolic compounds" and "bioactive metabolites of oleuropein," and olive oil mill wastewater, a by-product of olive oil production, would attenuate bone loss in ovariectomized rats. Of the 140 rats in this study, 120 had their ovaries surgically removed. After their procedures, the majority of these rats took supplemental tyrosol or hydroxytyrosol or olive oil mill wastewater or one of two different doses of olive oil mill wastewater extract. The researchers found that tyrosol, hydroxytyrosol, and olive oil mill wastewater extract increased bone formation and prevented bone loss. The researchers concluded that their finding "suggest the possible relevance of the dietary intake of olive oil, based on the capability of its major polyphenols, hydroxytyrosol and tyrosol, to lower the risk of inflammation-induced osteopenia in estrogen-deficient animals by their antioxidant activity."[6]

In a study published in 2013 in *Osteoporosis International*, researchers from eight countries wanted to learn if adherence to a Mediterranean diet, including olive oil, would reduce the incidence of hip fractures. It is well-known that hip fractures have the potential to impact the quality of life, especially of elders.

"Hip fractures are associated with considerable disability, loss of independence, diminished quality of life, and reduced survival."

The cohort consisted of 188,795 women and men who participated in the European Prospective Investigation into Cancer and nutrition study; at enrollment, the mean age was 48.6. They were followed for a median period of nine years; during that time, researchers recorded 802 incident hip fractures. The researchers found an inverse relationship between consumption of a Mediterranean diet and incidence of hip fractures. Thus, those who consumed a Mediterranean diet had fewer hip fractures. This was especially true for men as well as women who were 60 years old or older at recruitment. The researchers concluded that their prospective study "found evidence that closer adherence to MD [Mediterranean diet] appears to protect against hip fracture occurrence, particularly among men."[7]

A final study on the relationship between consumption of olive oil and bone health was published in 2011 in *Osteoporosis International*. In this investigation, researchers from Spain examined the effects of oleuropein on stem cells from human bone marrow. The researchers found that when they added oleuropein to human stem cells, the number of osteoblasts, or cells that make bone, increased. The researchers concluded that "the Mediterranean diet, which includes the consumptions of large quantities of olive oil, could contribute to decreasing the risk of developing osteoporosis."[8]

IS OLIVE OIL BENEFICIAL FOR BONE HEALTH?

Throughout the world, millions of people are dealing with osteopenia and osteoporosis. If adding olive oil to one's diet can reduce the incidence of these potentially serious medical problems, as the previously noted research indicates, then olive oil should be added to the diet of just about everyone.

NOTES

1. Kontogianni, Meropi D., Labros Melistas, Mary Yannakoulia et al. February 2009. "Association between Dietary Patterns and Indices of Bone Mass in a Sample of Mediterranean Women." *Nutrition* 25(2): 165–171.

2. Vignini, Arianna, Laura Nanetti, Francesca Raffaelli et al. 2008. "Effect of Supplementation with Fortified Olive Oil on Biochemical Markers of Bone Turnover in Healthy Women." *Mediterranean Journal of Nutrition and Metabolism* 1(2): 117–120.

3. Saleh, N.K. and H.A. Saleh. February 4, 2011. "Olive Oil Effectively Mitigates Ovariectomy-Induced Osteoporosis in Rats." *BMC Complementary & Alternative Medicine* 11(1): 10.

4. Fernández-Real, José Manuel, Mónica Bulló, José Maria Moreno-Navarrete et al. October 2012. "A Mediterranean Diet Enriched with Olive Oil Is Associated with Higher Serum Total Osteocalcin Levels in Elderly Men at High Cardiovascular Risk." *The Journal of Clinical Endocrinology and Metabolism* 97(10): 3792–3798.

5. Puel, Caroline, Jacinthe Mathey, Apostolis Agalias et al. 2006. "Dose-Response Study of Effect of Oleuropein, an Olive Oil Polyphenol, in an Ovariectomy/Inflammation Experimental Model of Bone Loss in the Rat." *Clinical Nutrition* 25(5): 859–868.

6. Puel, Caroline, Julie Mardon, Apostolis Agalias et al. 2008. "Major Phenolic Compounds in Olive Oil Modulate Bone Loss in an Ovariectomy/Inflammation Experimental Model." *Journal of Agricultural and Food Chemistry* 56(20): 9417–9422.

7. Benetou, V., P. Orfanos, U. Pettersson-Kymmer et al. May 2013. "Mediterranean Diet and Incidence of Hip Fractures in a European Cohort." *Osteoporosis International* 24(5): 1587–1598.

8. Santiago-Mora, R., A. Casada-Díaz, M. D. De Castro, and J. M. Quesada-Gómez. February 2011. "Oleuropein Enhances Osteoblastogenesis and Inhibits Adipogenesis: The Effect on Differentiation in Stem Cells Derived from Bone Marrow." *Osteoporosis International* 22(2): 675–684.

REFERENCES AND RESOURCES

Magazines, Journals, and Newspapers

Benetou, V., P. Orfanos, U. Pettersson-Kymmer et al. May 2013. "Mediterranean Diet and Incidence of Hip Fractures in a European Cohort." *Osteoporosis International* 24(5): 1587–1598.

Fernández-Real, José Manuel, Mónica Bulló, José Maria Moreno-Navarrete et al. October 2012. "A Mediterranean Diet Enriched with Olive Oil Is Associated with High Serum Total Osteocalcin Levels in Elderly Men at High Cardiovascular Risk." *The Journal of Clinical Endocrinology and Metabolism* 97(10): 3792–3798.

Kontogianni, Meropi D., Labros Melistas, Mary Yannakoulia et al. February 2009. "Association between Dietary Patterns and Indices of Bone Mass in a Sample of Mediterranean Women." *Nutrition* 25(2): 165–171.

Puel, Caroline, Jacinthe Mathey, Apostolis Agalias et al. 2006. "Dose-Response Study of Effect of Oleuropein, an Olive Oil Polyphenol, in an Ovariectomy/Inflammation Experimental Model of Bone Loss in the Rat." *Clinical Nutrition* 25(5): 859–868.

Puel, Caroline, Julie Mardon, Apostolis Agalias et al. 2008. "Major Phenolic Compounds in Olive Oil Modulate Bone Loss in an Ovariectomy/Inflammation Experimental Model." *Journal of Agricultural and Food Chemistry* 56(20): 9417–9422.

Saleh, N. K. and H. A. Saleh. February 4, 2011. "Olive Oil Effectively Mitigates Ovariectomy-Induced Osteoporosis in Rats." BMC *Complementary & Alternative Medicine* 11(1): 10.

Santiago-Mora, R., A. Casado-Díaz, M. D. De Castro, and J. M. Quesada-Gómez. February 2011. "Oleuropein Enhances Osteoblastogenesis and Inhibits Adipogenesis: The Effect on Differentiation in Stem Cells Derived from Bone Marrow." *Osteoporosis International* 22(2): 675–684.

Vignini, Arianna, Laura Nanetti, Francesca Raffaelli et al. 2008. "Effect of Supplementation with Fortified Olive Oil on Biochemical Markers of Bone Turnover in Healthy Women." *Mediterranean Journal of Nutrition and Metabolism* 1(2): 117–120.

Website

The George Mateljan Foundation. www.whfoods.com.

Olive Oil and Cancer

This entry examines studies related to olive oil's supposed cancer-fighting properties. Subsequent entries look at specific types of cancer more closely.

MAY BE USEFUL IN THE FIGHT AGAINST CANCER
AND TREATMENT COMPLICATIONS

In a randomized, controlled trial published in 2012 in *Pediatric Hematology and Oncology*, researchers from Egypt noted that it is not uncommon for people to develop oral mucositis, the breaking down of cells leading to ulceration and infection, in response to chemotherapy and radiation. Yet, they maintained that no one has created an effective treatment for this painful condition, "which may greatly complicate the management of cancer and compromise cure rates." As a result, the researchers decided to test the use of honey and a mixture of honey, olive oil propolis extract, and beeswax to treat chemotherapy-related oral mucositis.

The cohort consisted of 90 children between the ages of 2 and 18 who had acute lymphoblastic leukemia and chemotherapy-related oral mucositis. Every

Coratina olives flying out of a washer before being milled into oil at Round Pond in Rutherford, California. (AP Photo/Eric Risberg)

child was assigned to one of three groups; each group had 30 children. The children in the first group had their affected areas treated with honey; the children in the second group received topical treatments of honey, olive oil propolis extract, and beeswax; and the children in the third group, which served as controls, were treated with topical benzocaine. All of the children were treated for no more than 10 days.

The researchers found that the patients in all the groups experienced complete healing. However, there were differences in the amount of time required to heal. Both the honey and the honey combination groups produced faster healing that the benzocaine group. But, the honey had faster results than the combination product. The researchers recommended that future studies on treatments for chemotherapy-induced mucositis use "honey and possibly other bee products and olive oil."[1]

In a study published in 2012 in the *International Journal of Oncology*, researchers from Italy and Ohio (United States) noted that olive oil is filled with the polyphenol oleuropein. Because oleuropein has such strong antioxidant properties, they wondered if it would be useful against prostate cancer. The researchers decided to investigate using prostate cancer cell lines. The researchers found that when they treated prostate cancer cell lines with oleuropein for 72 hours, there was "a significant reduction in cell viability." Oleuropein "decreased prostate cancer cell proliferation and induced necrotic cell death."[2]

In a study published in 2009 in *Molecular Nutrition & Food Research*, researchers from Greece examined the ability of olive leaf extract to kill human breast and bladder cancer cells. The researchers found that the majority of the olive leaf extracts that were tested had "profound antioxidant activity." They also tested two of the olive leaf extracts for the ability to inhibit cancer cell growth. Both extracts "inhibited cell proliferation." The researchers noted that their findings "are in line with recently demonstrated cancer inhibitory effects of phytochemical derivatives of [the] olive tree."[3]

In a study published in 2009 in the *International Journal of Cancer*, researchers from many different European countries investigated the association between dietary factors and upper-aerodigestive tract cancer. (Aerodigestive is the combined organs and tissues of the respiratory tract and the upper part of the digestive tract.) The cohort consisted of 2,304 people with upper-aerodigestive tract cancer and 2,227 controls who were recruited from 14 centers in 10 European countries. The researchers found that the subjects living in Mediterranean countries consumed olive oil regularly; meanwhile, the subjects living in non-Mediterranean countries were infrequent users of the oil. The researchers observed a significant inverse association between the consumption of olive oil and risk of upper-aerodisgestive tract cancer. So, olive oil users had a reduced risk of upper-aerodigestive cancers. The researchers concluded that they "found strong evidence that dietary habits play an important role in the etiology of these cancers."[4]

In a study published in 2011 in the *European Journal of Cancer*, researchers from Belgium, the United Kingdom, and the Netherlands examined the effect of olive oil, animal products, and other major dietary fats on the risk of bladder cancer. The cohort consisted of 200 people who were diagnosed with bladder cancer and 386 healthy controls. All of the participants were sent a food frequency questionnaire to complete. Three trained interviewers visited the subjects and controls in their homes. The researchers found a statistically significant inverse association between the intake of olive oil and the risk for bladder cancer. Thus, a higher intake of olive oil reduced the risk of bladder cancer. The researchers concluded that they "observed a potentially protective effect of a high intake of olive oil."[5]

In a study published in 2013 in *Nutrition and Cancer*, researchers from Spain tested the ability of the unsaponifiable fraction of extra-virgin olive oil to prevent the spread and kill human colon cancer cells in the laboratory setting. After a number of different laboratory procedures, the researchers learned that the unsaponifiable fraction of extra-virgin olive oil had a strong ability to stop the growth and destroy human colon cancer cells. The researchers concluded that the unsaponifiable fraction of extra-virgin olive oil "may provide a basis for developing natural sources of chemopreventive and chemotherapeutic agents for the treatment of human colon cancer."[6]

In a case-controlled trial published in 2012 in the *Journal of the American College of Nutrition*, researchers from Canada, Iran, and Boston, Massachusetts (United States), examined the association between adherence to a Mediterranean-style diet and the risk of esophageal squamous cell cancer, a very lethal type of cancer that is somewhat common in Iran. The cohort consisted of 47 people with esophageal squamous cell cancer and 96 controls between the ages of 40 and 75. The researchers found that the Iranians who were more compliant with a Mediterranean-style dietary pattern, in which olive oil is the sole fat, "strongly and significantly" reduced their risk of esophageal squamous cell cancer. "The impact of olive oil intake in protection against several neoplasms [new and abnormal growths, especially characteristic of cancer], including esophageal carcinomas, has been shown previously and was confirmed in the present research."[7]

In a report published in 2011 in *Lipids in Health and Disease*, researchers from Athens, Greece conducted a meta-analysis of studies on the association between olive oil intake and the prevalence of cancer. The cohort consisted of 19 case-controlled studies, which included 13,800 people with cancer and 23,340 controls. After reviewing their data, the researchers found "olive oil consumption was associated with lower odds of cancer development." In fact, according to the researchers, people who had the highest intake of olive oil had a 34% lower risk of having any type of cancer. Still, the researchers learned that the benefits of olive oil appeared to be most significant for breast cancer and cancers of the digestive tract. The researchers noted that "the strength and consistency of the findings states a hypothesis about the protective role of olive oil intake on cancer risk."[8]

IS OLIVE OIL BENEFICIAL IN THE
FIGHT AGAINST CANCER?

The research reviewed for this entry indicates that olive oil provides a measure of support against cancer. People who regularly include olive oil in their diets appear to have lower rates of cancer. That seems to be a good reason to include moderate amounts of olive oil in the diet. Why only moderate amounts? It is important to remember that olive oil, like all other oils, has a high caloric count.

NOTES

1. Abdulrhman, Mamdouh, Nancy Samir El Barbary, Dina Ahmed Amin, and Rania Saeid Ebrahim. 2012. "Honey and a Mixture of Honey, Beeswax, and Olive Oil—Propolis Extract in Treatment of Chemotherapy-Induced Oral Mucositis: A Randomized Controlled Pilot Study." Pediatric Hematology and Oncology 29(3): 285–292.

2. Acquaviva, Rosaria, Claudia Di Giacomo, Valeria Sorrenti et al. 2012. "Antiproliferative Effect of Oleuropein in Prostate Cell Lines." International Journal of Oncology 41(1): 31–38.

3. Goulas, Vlassios, Vassiliki Exarchou, Anastassios N. Troganis et al. 2009. "Phytochemicals in Olive-Leaf Extracts and Their Antiproliferative Activity against Cancer and Endothelial Cells." Molecular Nutrition & Food Research 53(5): 600–608.

4. Lagiou, P., R. Talamini, E. Samoli et al. June 1, 2009. "Diet and Upper-Aerodigestive Tract Cancer in Europe: The ARCAGE Study." International Journal of Cancer 124(11): 2671–2676.

5. Brinkman, M. T., F. Buntinx, E. Kellen et al. February 2011. "Consumption of Animal Products, Olive Oil and Dietary Fat Results from the Belgian Case-Control Study on Bladder Cancer Risk." European Journal of Cancer 47(3): 436–442.

6. Cárdeno, A., M. Sánchez-Hidalgo, A. Cortes-Delgado, and C. Alarcón de la Lastra. 2013. "Mechanisms Involved in the Antiproliferative and Proapoptotic Effects of Unsaponifiable Fraction of Extra Virgin Olive Oil on HT-29 Cancer Cells." Nutrition and Cancer 65(6): 908–918.

7. Jessri, Mahsa, Bahram Rashidkhani, Bahareh Hajizadeh, and Paul F. Jacques. October 2012. "Adherence to Mediterranean-Style Dietary Pattern and Risk of Esophageal Squamous Cell Carcinoma: A Case-Control Study in Iran." Journal of the American College of Nutrition 31(5): 338–351.

8. Psaltopoulou, T., R. I. Kosti, D. Haidopoulos et al. July 30, 2011. "Olive Oil Intake Is Inversely Related to Cancer Prevalence: A Systematic Review and a Meta-Analysis of 13,800 Patients and 23,340 Controls in 19 Observational Studies." Lipids in Health and Disease 10: 127+.

REFERENCES AND RESOURCES

Magazines, Journals, and Newspapers

Abdulrhman, Mamdouh, Nancy Samir El Barbary, Dina Ahmed Amin, and Rania Saeid Ebrahim. 2012. "Honey and a Mixture of Honey, Beeswax, and Olive Oil—Propolis

Extract in Treatment of Chemotherapy-Induced Oral Mucositis: A Randomized Controlled Pilot Study." *Pediatric Hematology and Oncology* 29(3): 285–292.

Acquaviva, Rosaria, Claudia Di Giacomo, Valeria Sorrenti et al. 2012. "Antiproliferative Effect of Oleuropein in Prostate Cell Lines." *International Journal of Oncology* 41(1): 31–38.

Brinkman, M. T., F. Buntinx, E. Kellen et al. February 2011. "Consumption of Animal Products, Olive Oil and Dietary Fat and Results from the Belgian Case-Control Study on Bladder Cancer Risk." *European Journal of Cancer* 47(3): 436–442.

Cárdeno, A., M. Sánchez-Hidalgo, A. Cortes-Delgado, and C. Alarcón de la Lastra. 2013. "Mechanisms Involved in the Antiproliferative and Proapoptotic Effects of Unsaponifiable Fraction of Extra Virgin Olive Oil on HT-29 Cancer Cells." *Nutrition and Cancer* 65(6): 908–918.

Goulas, Vlassios, Vassiliki Exarchou, Anastassios N. Troganis et al. 2009. "Phytochemicals in Olive-Leaf Extracts and Their Antiproliferative Activity against Cancer and Endothelial Cells." *Molecular Nutrition & Food Research* 53(5): 600–608.

Jessri, Mahsa, Bahram Rashidkhani, Bahareh Hajizadeh, and Paul F. Jacques. October 2012. "Adherence to Mediterranean-Style Dietary Pattern and Risk of Esophageal Squamous Cell Carcinoma: A Case-Controlled Study in Iran." *Journal of the American College of Nutrition* 31(5): 338–351.

Lagiou, P., R. Talamini, E. Samoli et al. June 1, 2009. "Diet and Upper-Aerodigestive Tract Cancer in Europe: The ARCAGE Study." *International Journal of Cancer* 124(11): 2671–2676.

Psaltopoulou, T., R. I. Kosti, D. Haidopoulos et al. July 30, 2011. "Olive Oil Intake Is Inversely Related to Cancer Prevalence: A Systematic Review and Meta-Analysis of 13,800 Patients and 23,340 Controls in 19 Observational Studies." *Lipids in Health and Disease* 10: 127+.

Website

International Olive Council. www.internationaloliveoil.org.

Olive Oil and Breast Cancer

The previous entry explored olive oil's impact on cancer in general; this entry examines the purported benefits of olive oil in preventing and treating breast cancer specifically.

USEFUL FOR THE PREVENTION AND MAY BE THE TREATMENT OF BREAST CANCER

In a study published in 2009 in the *American Journal of Epidemiology*, researchers based in France examined the association between two types of dietary patterns—Western style and Mediterranean style and risk for developing breast cancer. The researchers included 2,381 postmenopausal women who were diagnosed with invasive breast cancer during a median follow-up period of 9.7 years. Among the foods included in the Western style diet were processed meats, French fries, rice/pasta, canned fish, cakes, butter, cream, and mayonnaise; this diet also included alcohol. The Mediterranean style diet was characterized by a high intake of fruits, vegetables, fresh fish, olives, olive oil, and sunflower oil. The researchers found a positive association between the Western style diet and risk for breast cancer. At the same time, the Mediterranean style diet was associated with a lower risk for breast cancer. However, the researchers added, these associations were limited to slim to normal weight women for the Western style diet and to women who had a low energy intake for the Mediterranean style diet. And, they concluded that the risk of breast cancer in postmenopausal women may be influenced by diet. "The avoidance of Western-type foods may reduce breast cancer risk in normal-weight women."[1]

In a study published in 2012 in BMC Cancer, researchers from Greece, the United Kingdom, and Italy evaluated whether the adherence of a Mediterranean diet pattern among Greek Cypriot women is associated with the risk of breast cancer. The cohort initially consisted of 1,109 women, between the ages of 40 and 70, with a confirmed diagnosis of breast cancer and 1,177 women controls. By the end of the study, there were 935 cases and 817 controls. After completing a number of different analyses, the researchers found that eating a diet pattern that is rich in olive oil, vegetables, fish, and legumes was inversely associated with the risk for breast cancer. Thus, a diet rich with these foods is linked to a reduced risk for cancer. Why is this important? "Considering the high incidence of bc [breast cancer] and assuming a causative association, even a small reduction in risk conferred by dietary components can be translated into several prevented cases per year, in Cyprus alone."[2]

In a study published in 2013 in the Journal of Medicinal Plants Research, researchers from Saudi Arabia tested the ability of hydroxytyrosol, a phenolic compound found in olive oil, to kill breast cancer cells in two different cell lines. The researchers found that hydroxytyrosol triggered cell death (apoptosis) in both cell lines. They concluded that their findings "support the hypothesis that hydroxytyrosol may exert a protective effect against breast cancer by reducing cell viability, arresting the cell cycle and inducing apoptosis in these cells."[3]

In a study published in 2011 in Nutrients, researchers from Spain tested the ability of hydroxytyrosol to protect against oxidative DNA damage in human breast cells. After a number of different tests, the researchers found that hydroxytyrosol demonstrated "radical savaging capacity" and a strong ability to prevent oxidative stress in normal breast cells. In so doing, it stopped "the initiation of a chain of reactions to transform normal cells into cancer cells." However, while hydroxytyrosol was able to help prevent the formation of breast cancer cells, according to this research, it was unable to "protect against breast cancer once developed."[4]

In a study published in 2011 in Molecular Nutrition & Food Research, researchers from Spain investigated the ability of hydroxytyrosol to treat breast tumors in rats. The researchers began by inducing breast tumors in 28 female

Sprague-Dawley rats; 10 of these rats were then treated with hydroxytyrosol five days per week for a total of six weeks. The researchers found that hydroxytyrosol inhibited breast cancer growth in rats to a degree similar to doxorubicin, an anticancer medication with some potentially serious side effects. The researchers concluded that their findings demonstrated that "hydroxytyrosol exerted anti-tumour properties in Sprague-Dawley rats with experimental mammary tumours, inhibiting the growth and cell proliferation."[5]

In a study published in 2009 in *Cytotechnology*, researchers from Japan examined the antiproliferative and apoptotic effects of hydroxytyrosol and oleuropein, another phenolic compound in olive oil, on human breast cancer cells. After completing a number of different laboratory tests, the researchers found that hydroxytyrosol and oleuropein inhibited cell proliferation and induced cell death in a dose-dependent manner. Moreover, "constant consumption of olive products such as leaf tea and oil may enhance these protective effects thus, suggesting their daily consumption is more effective than the infrequent intake."[6]

In a study published in 2014 in the *International Journal of Cancer*, researchers from Spain explained that high mammographic density has been identified as a primary risk factor for the development of breast cancer. Still, according to these researchers, there have been few studies on the association between high mammographic density and diet. Their cohort consisted of 3,548 peri- and postmenopausal women recruited from seven breast cancer–screening programs in Spain; their mean age was 56.2 years. The researchers learned that the women who had a higher intake of olive oil had a reduced risk of high mammographic density. The researchers concluded that their study "supported the protective role of olive oil vis-à-vis breast cancer." In addition, "the protective effect is evident in women who reported a high consumption [of olive oil], which goes to reinforce both the existence of a risk gradient, and the interest that lies in increasing the intake of this food, an essential constituent of the Mediterranean diet."[7]

In a study published in 2006 in *Public Health Nutrition*, researchers from Spain noted that the incidence of breast cancer and rates of deaths from breast cancer are higher in the Canary Islands, particularly the island of Gran Canaria, than in other areas of Spain. As a result, the researchers decided to examine the effect of the consumption of olive oil and other fats on the risk of breast cancer on the Canary Islands, "where legislative initiatives and marketing have promoted olive oil consumption in the past 20 years."

The cohort consisted of 326 Canary Islands women with histologically confirmed cases of breast cancer and 492 controls. The mean age of the women with breast cancer was 55.5 years; the mean age for the controls was 53.1 years. The researchers found an inverse linear association between olive oil and risk for breast cancer. The women who used the most olive oil were least likely to develop the disease. The researchers noted that their findings "support the protective role of olive oil suggested by other studies carried out among Mediterranean populations." And, they advised the promotion of "olive oil consumption through political, economical, and social strategies."[8]

In a study published in 2012 in *Asian Pacific Journal of Cancer Prevention*, researchers from Saudi Arabia and Egypt tested the ability of oleuropein to prevent the spread of breast cancer from its initial site in the breast to other parts of the body. The researchers found that oleuropein has the potential to "induce anti-metastatic effects on human breast cancer cells." And, they concluded that "treatment of breast cancer cells with oleuropein could help in prevention of cancer metastasis."[9]

IS OLIVE OIL A BENEFICIAL ADDITION IN THE FIGHT AGAINST BREAST CANCER?

From the research studies reviewed, olive oil appears to have a number of benefits for those who wish to prevent breast cancer. There is also the possibility that it may be useful to those who have already developed this medical problem.

NOTES

1. Cottet, V., M. Touvier, A. Fournier et al. November 15, 2009. "Postmenopausal Breast Cancer Risk and Dietary Patterns in the E3N-EPIC Prospective Cohort Study." *American Journal of Epidemiology* 170(10): 1257–1267.

2. Demetriou, C. A., Hadjisavvas, M. A. Loizidou et al. March 23, 2012. "The Mediterranean Dietary Pattern and Breast Cancer Risk in Greek-Cypriot Women: A Case-Control Study." *BMC Cancer* 12: 113+.

3. Elamin, Maha H., Zeinab K. Hassan, Sawsan A. Omer et al. August 25, 2013. "Apoptotic and Antiproliferative Activity of Olive Oil Hydroxytyrosol on Breast Cancer Cells." *Journal of Medicinal Plants Research* 7(32): 2420–2428.

4. Warleta, Fernando, Christina Sánchez Quesada, María Campos et al. October 2011. "Hydroxytyrosol Protects against Oxidative DNA Damage in Human Breast Cells." *Nutrients* 3(10): 839–857.

5. Granados-Principal, Sergio, Jose L. Quiles, Cesar Ramirez-Tortosa et al. May 2011. "Hydroxytyrosol Inhibits Growth and Cell Proliferation and Promotes High Expression of Sfrp4 in Rat Mammary Tumours." *Molecular Nutrition & Food Research* 55(Supplement 1): S117–S126.

6. Han, Junkyu, Terence P. N. Talorete, Parida Yamada, and Hiroko Isoda. 2009. "Anti-Proliferative and Apoptotic Effects of Oleuropein and Hydroxytyrosol on Human Breast Cancer MCF-7 Cells." *Cytotechnology* 59(1): 45–53.

7. García-Arenzana, N., E. M. Navarrete-Muñoz, V. Lope et al. 2014. "Calorie Intake, Olive Oil Consumptions and Mammographic Density among Spanish Women." *International Journal of Cancer* 134(8): 1916–1925.

8. García-Segovia, Purificación, Almudena Sánchez-Villegas, Jorge Doreste et al. 2006. "Olive Oil Consumption and Risk of Breast Cancer in the Canary Islands: A Population-Based Case-Control Study." *Public Health Nutrition* 9(1A): 163–167.

9. Hassan, Zeinab K., Maha H. Elamin, Maha H. Daghestani et al. 2012. "Oleuropein Induces Anti-Metastatic Effects in Breast Cancer." *Asian Pacific Journal of Cancer Prevention* 13(9): 4555–4559.

REFERENCES AND RESOURCES

Magazines, Journals, and Newspapers

Cottet, V., M. Touvier, A. Fournier et al. November 15, 2009. "Postmenopausal Breast Cancer Risk and Dietary Patterns in the E3N-EPIC Prospective Cohort Study." *American Journal of Epidemiology* 170(10): 1257–1267.

Demetriou, C. A., A. Hadjisavvas, M. A. Loizidou et al. March 23, 2012. "The Mediterranean Dietary Pattern and Breast Cancer Risk in Greek-Cypriot Women: A Case-Control Study." *BMC Cancer* 12: 113+.

Elamin, Maha H., Zeinab K. Hassan, Sawsan A. Omer et al. August 25, 2013. "Apoptotic and Antiproliferative Activity of Olive Oil Hydroxytyrosol on Breast Cancer Cells." *Journal of Medicinal Plants Research* 7(32): 2420–2428.

García-Arenzana, N., E. M. Navarrete-Muñoz, V. Lope et al. April 15, 2014. "Calorie Intake, Olive Oil Consumption and Mammographic Density among Spanish Women." *International Journal of Cancer* 134(8): 1916–1925.

García-Segovia, Purificación, Almudena Sánchez-Villegas, Jorge Doreste et al. 2006. "Olive Oil Consumption and Risk of Breast Cancer in the Canary Islands: A Population-Based Case-Control Study." *Public Health Nutrition* 9(1A): 163–167.

Granados-Principal, Sergio, Jose L. Quiles, Cesar Ramirez-Tortosa et al. May 2011. "Hydroxytyrosol Inhibits Growth and Cell Proliferation and Promises High Expression of Sfrp4 in Rat Mammary Tumours." *Molecular Nutrition & Food Research* 55(Supplement 1): S117–S126.

Han, Junkyu, Terence P. N. Talorete, Parida Yamada, and Hiroko Isoda. 2009. "Anti-Proliferative and Apoptotic Effects of Oleuropein and Hydroxytyrosol on Human Breast Cancer MCF-7 Cells." *Cytotechnology* 59(1): 45–53.

Hassan, Zeinab K., Maha H. Elamin, Maha H. Daghestani et al. 2012. "Oleuropein Induces Anti-Metastatic Effects in Breast Cancer." *Asian Pacific Journal of Cancer Prevention* 13(9): 4555–4559.

Warleta, Fernando, Christina Sánchez Quesada, María Campos et al. October 2011. "Hydroxytyrosol Protects against DNA Damage in Human Breast Cells." *Nutrients* 3(10): 839–857.

Website

Olive Oil Source. www.oliveoilsource.com

Olive Oil and Colorectal Cancer

Olive oil may help prevent and counteract colorectal cancer.

MAY PLAY A ROLE IN PREVENTING AND TREATING COLORECTAL CANCER

In a study published in 2008 in the *International Journal of Cancer*, researchers from Ireland, Northern Ireland, Italy, and Finland conducted laboratory tests to determine if different doses of phenolics (hydroxytyrosol, tyrosol, pinoresinol, and caffeic acid) extracted from virgin olive oil could kill human colon cancer cells. After a number of different analyses, the researchers found that the phenolics from olive oil had "different dose-related anti-invasive effects." The researchers concluded that "phenols from virgin olive oil have the ability to inhibit invasion of colon cancer cells and the effects may be mediated at different levels of the invasion cascade."[1]

In a similar study published in 2009 in *Molecular Nutrition & Food Research*, researchers from the United Kingdom and Italy tested the ability of the hydroxytyrosol found in olive oil to prevent the spread of human colon cancer cells. The researchers exposed human colon cancer cells to either hydroxytyrosol or a vehicle (in this case 2% methanol). The researchers found that exposure to hydroxytyrosol resulted in "significant growth inhibition" of the human colon cells. And, this inhibition was observed in all of the concentrations that they tested. According to the researchers, this may help to explain "the inverse link between colon cancer and olive oil consumption."[2] So, these researchers contend that the more olive oil people consume, the less likely that that they will develop colon cancer.

In still another laboratory study on human colon cancer cells that was published in 2012 in the *Journal of Pharmaceutical and Biomedical Analysis*, researchers from Spain tested the antiproliferative activity of olive oil extracts. The researchers found that the amount of human colon cancer cell growth inhibition varied from extract to extract and from the different concentration levels of the extracts. "Thus, while concentrations of 0.1% reduced cell growth by up to 20% with respect to control of untreated cells in both cell lines, inhibition of cell proliferation of 0.01% was much less drastic." Moreover, there were "significant differences in growth inhibition induced by some of the 14 olive-oil samples."[3]

In a laboratory study published in 2008 in *Carcinogenesis*, researchers from Italy and Dallas, Texas (United States) wanted to assess the anticolon cancer properties of phenolic extracts in two different types of extra-virgin olive oil. Pinoresinol was the primary phenol in extra-virgin oil A, and oleocanthal was the primary phenol in extra-virgin oil B. The researchers evaluated the activity of extra-virgin olive oil on "cell proliferation, apoptosis and cell cycle regulation." The researchers found that the A oil with pinoresinol decreased cell viability, induced apoptosis, and "modulated cell cycle dynamics in the CRC [colorectal cancer] cell lines." On the other hand, the B oil had a "relatively weak ability" to have an effect on the cancer cells.[4]

In a laboratory study published in 2011 in *Genes & Nutrition*, researchers from Italy examined the ability of hydroxytyrosol and oleuropein, the primary polyphenols in olive oil, to destroy and prevent the spread of two types of human colorectal cancer cells. After a number of different tests, researchers learned that hydroxytyrosol was able to kill and stop the spread of both types of human colorectal cancer cells. On the other hand, oleuropein had an antiproliferative effect only on one type of human colorectal cancer cells and a "slight pro-apoptotic effect" in both cell lines. And, these changes "were observed at polyphenol concentrations corresponding to those daily consumed in some Mediterranean areas."[5]

In a study published in 2010 in *Clinical Nutrition*, researchers from Spain noted that people with inflammatory bowel disease are at increased risk for developing ulcerative colitis—associated colorectal cancer. As a result, they decided to test their hypothesis that extra-virgin olive oil, with its "anti-inflammatory, antiproliferative, and antiapoptotic effects," might lower that risk. The researchers began by dividing 84 female mice into two groups. One group was fed a diet consisting of 10% extra-virgin olive oil and the other group had a diet with 10% sunflower oil. To induce chronic ulcerative colitis, the mice were exposed to 15 cycles of 0.7% dextran sodium sulphate, which was administered in their drinking water. The mice were then divided again into smaller groups. The trial continued for six weeks. After the mice were sacrificed, the researchers determined that there was significantly more "disease activity" in the mice fed sunflower oil than the mice fed extra-virgin olive oil. They concluded that a diet rich in extra-virgin olive oil helps prevent the development of abnormal colorectal cells and colorectal cancer. "Olive oil appears as an example of a functional food with a variety range of constituents that could contribute to its overall chemopreventive/beneficial effects."[6]

In a study published a few years earlier, in 2006, in *Nutrition*, researchers from Pittsburgh, Pennsylvania (United States) investigated the effects of corn, olive or fish oil on the severity of chronic colitis and the development of colitis-associated premalignant changes in mice. At the age of eight weeks, mice that were bred to develop inflammation of the digestive tract were randomized into one of the three groups of the supplemental oil. The trial continued for 12 weeks. During that time, the mortality rate of the mice fed fish oil was significantly higher than the mice fed olive oil or corn oil. The mice fed fish oil also had cases of colitis that were more severe than the mice fed the other oils. After the mice were sacrificed, the researchers learned that the mice on olive oil had significantly less dysplasia than the mice taking corn oil and fish oil. In so doing, olive oil "decreased the risk of neoplasia associated with chronic colitis."[7]

In a study published in 2013 in *Anticancer Research*, researchers from Italy randomly divided mice that were bred to be at high risk for developing colon cancer into four groups. While one group of mice were just fed a regular mouse diet, the mice in the other three groups received supplemental olive or salmon or evening primrose oil. The trial continued for 10 weeks. The researchers found that the mice in all three of the treated groups experienced "a reduction in total intestinal polyp number and load." And, the treatments were safe. "None of the animals fed with experimental diets showed any observable toxicity or any gross changes attributable to liver, kidney or lung toxicity."[8]

In a study published in 2007 in the *Annals of Oncology*, researchers from Italy, France, and Switzerland investigated the relationship between fried foods, including foods fried in olive oil, and the risk of colorectal cancer. The data, which were obtained from a large multicenter case-control study of colorectal cancer, included 1,394 cases of colon cancer, 886 cases of rectal cancer, and 4,765 controls. The researchers did not find an association between consumption of fried foods and the incidence of colorectal cancer. However, when the researchers evaluated the specific fats used for frying, they "found that olive oil, but not other types of oils, appeared to protect from colon cancer." But, this effect was not also seen in rectal cancer.[9]

Very different results were obtained in a study published in 2008 in the *European Journal of Nutrition*. Researchers from Italy divided their rats into three groups. During the trial, the rats in each of the groups consumed diets containing 23% of lipids from extra-virgin olive oil rich in phenolic compounds or extra-virgin olive oil without the phenolic compounds or corn oil, which served as the control. The researchers used a chemical, known as 1,2-dimethylhydrazine, to induce colon cancer in the rats. Thirteen weeks after this chemical was first administered, some rats from each of the groups were sacrificed; 32 weeks after the first administration of the chemical, all of the rats were sacrificed. The researchers found no evidence that the olive oil had any effect on the development of colon cancer. According to the researchers, they found "a similar incidence, multiplicity and differentiation of tumours in all three dietary groups." And, they concluded that their "data do not support a protective role of olive oil against colon carcinogenesis."[10]

IS OLIVE OIL BENEFICIAL IN PREVENTING AND TREATING COLORECTAL CANCER?

A strong majority of the research on this topic concludes that olive oil is useful in both the prevention and treatment of colorectal cancer. There are a few studies that have found that olive oil is useless for this condition. Still, since olive oil has so many other benefits, it certainly is a good idea to include it in the diet.

NOTES

1. Hashim, Y. Z., I. R. Rowland, H. McGlynn et al. February 1, 2008. "Inhibitory Effects of Olive Oil Phenolics on Invasion of Human Colon Adenocarcinoma Cells *In Vitro*." *International Journal of Cancer* 122(3): 495–500.

2. Corona, G., M. Deiana, A. Incani et al. July 2009. "Hydroxytyrosol Inhibits the Proliferation of Human Colon Adenocarcinoma Cells through Inhibition of ERK1/2 and Cyclin D1." *Molecular Nutrition & Food Research* 53(7): 897–903.

3. Fernández-Arroyo, S., A. Gómez-Martínez, L. Rocamora-Reverte et al. 2012. "Application of nanoLC-ESI-TOF-MS for the Metabolomic Analysis of Phenolic Compounds from Extra-Virgin Olive Oil in Treated Colon-Cancer Cells." *Journal of Pharmaceutical and Biomedical Analysis* 63: 128–134.

4. Fini, Lucia, Erin Hotchkiss, Vincenzo Fogliano et al. 2008. "Chemopreventive Properties of Pinoresinol-Rich Olive Oil Involve a Selective Activation of the ATM-p53 Cascade in Colon Cancer Cell Lines." *Carcinogenesis* 29(1): 139–146.

5. Notarnicola, Maria, Simona Pisanti, Valeria Tutino et al. 2011. "Effects of Olive Oil Polyphenols on Fatty Acid Synthase Gene Expression and Activity in Human Colorectal Cancer Cells." *Genes & Nutrition* 6(1): 63–69.

6. Sánchez-Fidalgo, S., I. Villagas, A. Cárdeno et al. 2010. "Extra-Virgin Olive Oil-Enriched Diet Modulates DSS-Colitis-Associated Colon Carcinogenesis in Mice." *Clinical Nutrition* 29(5): 663–673.

7. Hegazi, Refaat A. F., Reda S. Saad, Hussam Mady et al. 2006. "Dietary Fatty Acids Modulate Chronic Colitis, Colitis-Associated Colon Neoplasia and COX-2 Expression in IL-10 Knockout Mice." *Nutrition* 22(3): 275–282.

8. Notarnicola, Maria, Valeria Tutino, Angela Tafaro et al. September 2013. "Antitumorigenic Effect of Dietary Natural Compounds *via* Lipid Metabolism Modulation in Apc(Min/+) Mice." *Anticancer Research* 33(9): 3739–3744.

9. Galeone, C., R. Talamini, F. Levi et al. January 2007. "Fried Foods, Olive Oil and Colorectal Cancer." *Annals of Oncology* 18(1): 36–39.

10. Femia, A. P., P. Dolara, M. Servili et al. September 2008. "No Effects of Olive Oils with Different Phenolic Content Compared to Corn Oil on 1,2-Dimethylhdrazine-Induced Colon Carcinogenesis in Rats." *European Journal of Nutrition* 47(6): 329–334.

REFERENCES AND RESOURCES

Magazines, Journals, and Newspapers

Corona, G., M. Deiana, A. Incani et al. July 2009. "Hydroxytyrosol Inhibits the Proliferation of Human Colon Adenocarcinoma Cells through Inhibition of ERK1/2 and Cyclin D1." *Molecular Nutrition & Food Research* 53(7): 897–903.

Femia, A. P., P. Dolara, M. Servili et al. September 2008. "No Effects of Olive Oils with Different Phenolic Content Compared to Corn Oil on 1, 2-Dimethylhdrazine-Induced Colon Carcinogenesis in Rats." *European Journal of Nutrition* 47(6): 329–334.

Fernández-Arroyo, S., A. Gómez-Martinez, L. Rocamora-Reverte et al. April 7, 2012. "Application of nanoLC-ESI-TOF-MS for the Metabolomic Analysis of Phenolic Compounds from Extra-Virgin Olive Oil in Treated Colon-Cancer Cells." *Journal of Pharmaceutical and Biomedical Analysis* 63: 128–134.

Fini, Lucia, Erin Hotchkiss, Vincenzo Fogliano et al. 2008. "Chemoprotective Properties of Pinoresinol-Rich Olive Oil Involve a Selective Activation of the ATM-p53 Cascade in Colon Cancer Cell Lines." *Carcinogenesis* 29(1): 139–146.

Galeone, C., R. Talamini, F. Levi et al. 2007. "Fried Foods, Olive Oil and Colorectal Cancer." *Annals of Oncology* 18(1): 36–39.

Hashim, Y. Z., I. R. Rowland, H. McGlynn et al. February 1, 2008. "Inhibitory Effects of Olive Oil Phenolics on Invasion of Human Colon Adenocarcinoma Cells *In Vitro*." *International Journal of Cancer* 122(3): 495–500.

Hegazi, Refaat A. F., Reda S. Sand, Hussam Mady et al. 2006. "Dietary Fatty Acids Modulate Colitis, Colitis-Associated Colon Neoplasia and COX-2 Expression in IL-10 Kockout Mice." *Nutrition* 22(3): 275–282.

Notarnicola, Maria, Simona Pisanti, Valeria Tutino et al. 2011. "Effects of Olive Oil Poly-
phenols on Fatty Acid Synthase Gene Expression and Activity in Human Colorectal
Caner Cells." *Genes & Nutrition* 6(1): 63–69.
Notarnicola, Maria, Valeria Tutino, Angela Tafaro et al. September 2013. "Antitumori-
genic Effect of Dietary Natural Compounds *via* Lipid Metabolism Modulation in
Apc(Min/+) Mice." *Anticancer Research* 33(9): 3739–3744.
Sánchez-Fidalgo, S., I. Villegas, A. Cárdeno et al. 2010. "Extra-Virgin Olive Oil-Enriched
Diet Modulates DSS-Colitis-Associated Colon Carcinogenesis in Mice." *Clinical Nu-
trition* 29(5): 663–673.

Website

International Olive Council. www.internationaloliveoil.org.

Olive Oil and Prostate Cancer

This entry examines olive oil's potential for preventing and treating prostate cancer.

USEFUL FOR THE PREVENTION AND MAY BE
A TREATMENT FOR PROSTATE CANCER

In a study published in 2012 in the *International Journal of Oncology*, research-ers from Toledo, Ohio (United States) and Italy tested the ability of oleuropein, a phytochemical found in olive oil, to kill prostate cancer cells in a laboratory set-ting. They also examined the effect of oleuropein on normal prostate cells. After a number of different studies, the researchers found that "oleuropein decreased prostate cancer cell proliferation and induced necrotic cell death." Moreover, oleuropein was able to differentiate between cancer cells and normal cells. The researchers concluded that "oleuropein treatment could be useful both for the prevention of tumour progression and/or therapy, either alone or in combination with conventional preventive or therapeutic agents, for the prevention and/or treatment of prostate cancer."[1]

In another laboratory study published in 2013 in *Current Cancer Drug Targets*, researchers from China examined the ability of hydroxytyrosol, a phytochemical found in olive oil, to destroy human prostate cancer cells and prevent them from spreading. The researchers found that hydroxytyrosol killed and "reduced the vi-ability of human prostate cancer cells." They noted that their findings "presented preliminary evidence on the *in vitro* chemopreventive effect of hydroxytyrosol and . . . contributed to further investigation of hydroxytyrosol as an anti-cancer agent."[2]

In a study published in 2013 in the *British Journal of Nutrition*, researchers from Korea investigated the use of maslinic acid, which is found in olive oil, to inhibit the spread of human prostate cancer cells. Interestingly, the concentration of maslinic acid in olive oil increases as the quality of the olive oil decreases.

In their report, the researchers explained that the spreading or "physical translocation" of cancer cells is actually a two-part process. First, the cells must break away from their primary location and move to another organ. Then, they need to settle into their new location. From their laboratory studies, the researchers learned that maslinic acid "inhibited the migration, invasion, and adhesion" of the cancerous human prostate cells. They concluded that maslinic acid may prove to be a useful agent and metastasis inhibitor.[3]

In a study published in 2013 in *JAMA Internal Medicine*, researchers from San Francisco, California, and Boston, Massachusetts (both in the United States) examined the dietary habits, including the intake of olive oil, among more than 4,500 men who were diagnosed with nonmetastatic prostate cancer, or prostate cancer that had not spread beyond the walnut-sized prostate gland. Careful reviews were conducted on the diets of all of the men. The researchers found that during the median follow-up period of 8.4 years, the men who replaced 10% of their total daily calories from carbohydrates, such as bread and rice, with vegetable fats, such as olive oil, had a 29% reduced risk of developing "lethal prostate cancer" (or prostate cancer that had spread) and a 26% reduced risk of dying. Moreover, the "men who consumed more vegetable fat after diagnosis had a lower risk of all-cause mortality."[4]

In a population-based case-control study published in 2004 in *Cancer Causes and Control*, researchers from Australia, New Zealand, and Italy wanted to learn which foods decreased and increased the risk of developing prostate cancer. The cohort consisted of 858 men who were younger than 70 years when they were diagnosed with prostate cancer. Their dietary intake was determined with a 121-item food frequency questionnaire. The researchers found that olive oil, which is commonly consumed with tomatoes, onions, and garlic, "could be beneficial with respect to prostate cancer."[5]

In 2009, *Molecular Nutrition & Food Research* published an article entitled "Can the Mediterranean Diet Prevent Prostate Cancer?" which was authored by researchers from Australia. These researchers noted that the Mediterranean diet contains "olive oil as the main added fat" as well as a good deal of fresh fruits, vegetables, and nuts. The researchers noted that Greek men who migrated to Australia more than half a century ago have continued to eat their Mediterranean diet. In sharp contrast to men born in Australia, these immigrants have a low rate of prostate cancer. "These migrants appear to have resisted acculturation and retained many important features of a traditional Mediterranean-style eating pattern, which may protect them from prostate cancer." The researchers concluded that the foods that make up the Mediterranean diet, such as olive oil, fruits and vegetables, "offer a palatable chemo-protective alternative for the prevention of prostate cancer."[6]

However, there are some research studies that have found evidence to support the hypothesis that olive oil neither prevents nor supports the formation of prostate cancer. For example, in a study published in 2008 in *the American Journal of Clinical Nutrition,* numerous researchers from multiple locations throughout the world wanted to learn if the intake of fat, including olive oil, had any effect on the incidence of prostate cancer. The researchers were able to obtain complete data and follow-up information on 142,520 men. Following a median follow-up of 8.7 years, prostate cancer was diagnosed in 2,727 men. The researchers found no significant association between dietary fat ("total, saturated, monounsaturated, and polyunsaturated and the ratio of polyunsaturated to saturated fat") and the risk of prostate cancer. And, they concluded that "the results from this large prospective study showed that a high intake of dietary fat was not associated with an increased risk of prostate cancer."[7]

In a study published in 2007 in *Cancer Causes & Control,* researchers from Sweden and New York City investigated the association between the intake of various types of fat and the risk of developing prostate cancer. The initial cohort consisted of 10,564 men between the ages of 45 and 73 years who were cancer-free. During a mean follow-up of 11 years, 817 cases of prostate cancer were diagnosed. Of these, 281 were classified as "advanced." And, 202 of the cases occurred before the age of 65 years. After a number of analyses, the researchers found "no association between intake of total, saturated or monounsaturated fat and risk of prostate cancer."[8]

In an article published in 2007 in *Minerva Urologica e Nefrologica,* researchers from Greece analyzed studies on the association between monounsaturated, polyunsaturated, and saturated fats and prostate cancer. In addition, they examined studies on the relationship between the Mediterranean diet and reduced risk for prostate cancer. Yet, they were unable to find definitive conclusions. They did acknowledge that the Mediterranean diet may be useful against prostate cancer growth. "It seems plausible," they wrote, "that factors associated with the traditional Mediterranean diet may have important inhibitory roles in prostate cancer development and progression." At the same time, they found it hard to determine which elements of the Mediterranean diet play this role. Moreover, according to these researchers, "the role of dietary fat seems to be those of growth enhancer, rather than cancer initiator, while associations between dietary fat component of Mediterranean diet and prostate cancer risk remains unclear." They suggested that it is possible that the anticancer effect may be a result of another component of the Mediterranean diet, such as plant foods or wine.[9]

IS OLIVE OIL USEFUL IN THE PREVENTION AND TREATMENT OF PROSTATE CANCER?

The results for prostate cancer are not a definitive as those for other medical problems. Still, the use of olive oil is so beneficial for other medical problems that it should probably be included in the diets of just about everyone. Hopefully, future research will yield more findings on the association between olive oil and prostate cancer prevention and treatment.

NOTES

1. Acquaviva, Rosaria, Claudia Di Giacomo, Valeria Sorrenti et al. 2012. "Antiproliferative Effect of Oleuropein in Prostate Cell Lines." *International Journal of Oncology* 41(1): 31–38.

2. Luo, C., Y. Li, H. Wang et al. July 2013. "Hydroxytyrosol Promotes Superoxide Production and Defects in Autophagy Leading to Anti-Proliferation and Apoptosis on Human Prostate Cancer Cells." *Current Cancer Drug Targets* 13(6): 625–639.

3. Park, S. Y., C. W. Nho, D. Y. Kwon et al. January 28, 2013. "Maslinic Acid Inhibits the Metastatic Capacity of DU145 Human Prostate Cancer Cells: Possible Mediation via Hypoxia-Inducible Factor-1α Signalling." *British Journal of Nutrition* 109(2): 210–222.

4. Richman, Erin L., Stacey A. Kenfield, Jorge E. Chavarro et al. 2013. "Fat Intake after Diagnosis and Risk of Lethal Prostate Cancer and All-Cause Mortality." *JAMA Internal Medicine* 173(14): 1318–1326.

5. Hodge, Allison M., Dallas R. English, Margaret R. E. McCredie et al. 2004. "Food, Nutrients and Prostate Cancer." *Cancer Causes and Control* 15(1): 11–20.

6. Itsiopoulos, Catherine, Allison Hodge, and Mary Kaimakamis. 2009. "Can the Mediterranean Diet Prevent Prostate Cancer." *Molecular Nutrition & Food Research* 53(2): 227–239.

7. Crowe, Francesca L., Timothy J. Key, Paul N. Appleby et al. May 2008. "Dietary Fat Intake and Risk of Prostate Cancer in the European Prospective Investigation into Cancer and Nutrition." *The American Journal of Clinical Nutrition* 87(5): 1405–1413.

8. Wallström, Anders Bjartell, Bo Gullberg et al. 2007. "A Prospective Study on Dietary Far and Incidence of Prostate Cancer (Malmö, Sweden)." *Cancer Causes & Control* 18(10): 1107–1121.

9. Stamatiou, K., D. Delakas, and F. Sofras. March 2007. "Mediterranean Diet, Monounsaturated: Saturated Fat Ratio and Low Prostate Cancer Risk. A Myth or a Reality?" *Minerva Urologica e Nefrologica* 59(1): 59–66.

REFERENCES AND RESOURCES

Magazines, Journals, and Newspapers

Acquaviva, Rosaria, Claudia Di Giacomo, Valeria Sorrenti et al. July 2012. "Antiproliferative Effect of Oleuropein in Prostate Cell Lines." *International Journal of Oncology* 41(1): 31–38.

Crowe, Francesca L., Timothy J. Key, Paul N. Appleby et al. May 2008. "Dietary Fat Intake and Risk of Prostate Cancer in the European Prospective Investigation into Cancer and Nutrition." *The American Journal of Clinical Nutrition* 87(5): 1405–1413.

Hodge, Allison M., Dallas R. English, Margaret R. E. McCredie et al. 2004. "Foods, Nutrients, and Prostate Cancer." *Cancer Causes and Control* 15(1): 11–20.

Itsiopoulos, Catherine, Allison Hodge, and Mary Kaimakamis. 2009. "Can the Mediterranean Diet Prevent Prostate Cancer?" *Molecular Nutrition & Food Research* 53(2): 227–239.

Luo, C., Y. Li, H. Wang et al. July 2013. "Hydroxytyrosol Promotes Superoxide Production and Defects in Autophagy Leading to Anti-Proliferation and Apoptosis on Human Prostate Cancer Cells." *Current Cancer Drug Targets* 13(6): 625–639.

Park, S. Y., C. W. Nho, D. Y. Kwon et al. January 28, 2013. "Maslinic Acid Inhibits the Metastatic Capacity of DU145 Human Prostate Cancer Cells: Possible Mediation via Hypoxia-Inducible Factor-1α Signalling." *British Journal of Nutrition* 109(2): 210–222.

Richman, Erin L., Atacey A. Kenfield, Jorge E. Chavarro et al. July 22, 2013. "Fat Intake after Diagnosis and Risk of Lethal Prostate Cancer and All-Cause Mortality." *JAMA Internal Medicine* 173(14): 1318–1326.

Stamatiou, K., D. Delakas, and F. Sofras. March 2007. "Mediterranean Diet, Monounsaturated: Saturated Fat Ratio and Low Prostate Cancer Risk. A Myth or a Reality?" *Minerva Urologica e Nefrologica* 59(1): 59–66.

Wallström, Peter, Anders Bjartell, Bo Gullberg et al. 2007. "A Prospective Study on Dietary Fat and Incidence of Prostate Cancer (Malmö, Sweden)." *Cancer Causes & Control* 18(10): 1107–1121.

Website

Olive Oil Times. www.oliveoiltimes.com.

Olive Oil and Cardiovascular Health

This entry discusses olive oil's potential for improving cardiovascular health.

SUPPORTS CARDIOVASCULAR HEALTH

In an article published in 2011 in *Maturitas*, researchers from Spain noted that the incidence of worldwide deaths from cardiovascular disease has been rising rapidly. Future projections indicate that this trend will continue. "Worldwide deaths from cardiovascular disease are projected to rise from 17.1 million in 2004 to 23.4 million in 2030." At the same time, there are lower rates of cardiovascular-related deaths in countries such as Spain and Greece where people tend to eat a Mediterranean diet that includes olive oil. The researchers described a hospital-based, case-control study that was conducted at the University of Navarra, Spain. The cohort consisted of 171 people who suffered their first heart attack (acute myocardial infarction) and 171 age, gender, and hospital-matched controls. Not surprisingly, the researchers found that "olive oil consumption may substantially reduce the risk of coronary disease."[1]

In a crossover study published in 2011 in *Nutrition, Metabolism & Cardio-vascular Diseases*, researchers from Spain and Brazil recruited 14 women and 12 men with elevated levels of cholesterol to follow a generally healthful diet for four weeks. Then, during three separate four-week periods, the volunteers ate a Mediterranean-style diet supplemented with virgin olive oil or walnuts or almonds. The researchers supplied the virgin olive oil, walnuts, and almonds. There were no washout periods between the diets. The researchers obtained fasting blood samples at baseline and at the end of each dieting period. Nine women and nine men completed the study. The researchers found that all three diets significantly lowered levels of total cholesterol, low-density lipoprotein (LDL or "bad" cholesterol), and the LDL/HDL ratio. (HDL is high-density lipoprotein cholesterol or "good" cholesterol.) These are all indicators of improved cardiovascular health. And, the researchers concluded that virgin olive oil, walnuts, and almonds, "which are foods rich in unsaturated fatty acids and other bioactive nutrients, should be included in healthy diets, especially those aimed at reducing blood cholesterol and cardiovascular risk."[2]

In a study published in 2011 in *Neurology*, researchers from France wanted to learn if higher intakes of olive oil were associated with lower incidences of stroke in older subjects. During the study, the researchers reviewed the medical records of 7,625 people, 65 years and older, from three cities in France. No one included in the study had a stroke history. There were three categories of olive oil consumption—no use, moderate use (in cooking or as a dressing or with bread), or intensive use (use both for cooking and dressing and with bread). After a little more than five years, 148 of the participants had strokes. When compared to those who never used olive oil, the researchers found that those with an intensive use of olive oil had a 41% lower risk of stroke. The researchers noted that "the high prevalence of stroke in older subjects emphasizes the need for primary and secondary prevention in this age group." So, their study, which demonstrated "a strong association between intensive olive oil use and lower stroke incidence . . . suggests a novel approach of dietary recommendations to prevent stroke occurrence in elderly populations."[3]

In a study published in 2011 in *The American Journal of Clinical Nutrition*, researchers from Italy and the United Kingdom investigated the association between consumption of fruit, vegetables, and olive oil and the incidence of coronary heart disease. The cohort consisted of 29,689 pre- and postmenopausal women enrolled in the European Prospective Investigation into Cancer and Nutrition (EPIC) group. During the average follow-up of 7.85 years, the researchers identified 144 major coronary heart disease events; most of these were nonfatal heart attacks. The researchers found that the consumption of olive oil reduced the risk of cardiovascular problems. They concluded that "an inverse association between increasing consumption of leafy vegetables and olive oil and CHD [coronary heart disease] risk emerged in this large cohort of Italian women."[4]

In a study published in 2013 in the *European Journal of Clinical Nutrition*, researchers from many different locations in Spain noted that cardiovascular disease "is the leading cause of mortality, accounting for 31.1% of all deaths." That is why the researchers decided to examine the association "between consumption of olive oil and the presence of cadiometabolic risk factors" in 4,572 residents of Spain who were 18 years or older. The researchers learned that over 90% of their cohort used olive oil for cooking and dressing and 68.9% for frying. Olive oil users tended to have lower levels of body mass index (BMI), triglycerides, LDL, and they were less likely to be obese. The researchers concluded that their study demonstrated that the consumption of olive oil "has a beneficial effect on these cardiovascular risk factors, particularly in the presence of obesity, impaired glucose regulation or a sedentary lifestyle, reducing the synergy between the different cardiovascular risk factors."[5]

In a double-blind, randomized trial published in 2013 in the *European Journal of Nutrition*, researchers from Rochester, Minnesota (United States), and Italy wanted to learn if olive oil would be useful for people with the early stages of atherosclerosis (accumulation of plaque in the arteries). The initial cohort consisted of 82 patients. For four months, their daily diets included 30mL of olive oil or 30mL of olive oil supplemented with EGCG (epigallocatechin-3-gallate), a green tea–derived phenol. Fifty-two subjects completed the entire trial. The researchers found that both groups experienced reductions in the accumulation of plaque formation; the members of the group taking olive oil with EGCG improved about the same as the members of the group on olive oil alone. According to the researchers, "this is the first such demonstration of such a permanent endothelial benefit via long-term supplementation of a macronutrient." And, they concluded that "supplementation with OO [olive oil] seems a reasonably easy and relatively cheap dietary measure to improve the endothelial function and perhaps favorably alter the progression of atherosclerotic disease, particularly in patients with already markedly impaired endothelial function."[6]

In a parallel-group, multicenter, randomized trial published in 2013 in *The New England Journal of Medicine*, researchers from Spain recruited 7,447 people, between the ages of 55 and 80. All of the participants had an increased risk for cardiovascular disease but were without cardiovascular disease when enrolled. Increased risk could be caused by a number of different factors such as a type 2 diabetes diagnosis. The participants were place on one of three diets—a Mediterranean diet supplemented with approximately 1L/week of extra-virgin olive oil, a Mediterranean diet supplemented with 30g/day of mixed nuts, or a control diet, consisting of advice on reducing dietary fat. The median follow-up time was 4.8 years. Although the researchers were unable to follow more than 500 participants, the results they obtained from the remaining subjects were notable. Adhering to the Mediterranean diet supplemented with olive oil or nuts significantly reduced the incidence of major cardiovascular events. "An energy-unrestricted Mediterranean diet supplemented with either extra-virgin olive oil or nuts results in an absolute reduction of approximately 3 major cardiovascular events per 1000 person-years, for a relative risk reduction of approximately 30%, among high-risk persons who were initially free of cardiovascular disease."[7]

In a study published in 2012 in the *British Journal of Nutrition*, researchers from Spain examined the association between consumption of olive oil and coronary heart disease in the EPIC Spanish cohort study. The cohort consisted of 40,142 subjects, of whom 38% were males. At recruitment, they were all free of coronary heart disease events. During a multiyear follow-up period, 587 coronary heart disease events were recorded. The researchers found an inverse association between the consumption of olive oil and coronary heart disease. And, virgin olive oil was more effective in the prevention of coronary heart disease than ordinary olive oil. The researchers concluded that their study "supports the contribution that olive oil makes within this dietary pattern in terms of reducing risk of CHD [coronary heart disease]."[8]

In a double-blind, randomized, crossover trial published in 2012 in the *American Journal of Hypertension*, researchers from Spain examined the effect that the intake of olive oil has on the blood pressure and endothelial function. The cohort consisted of 24 young women with high–normal blood pressure or stage 1 essential hypertension; their average age was 26 years. The women were placed on diets with polyphenol olive oil or diets with polyphenol-free olive oil. Each dietary period continued for two months; and there was a four-week washout period between the two diets. The researchers found that the diets with polyphenol olive oil lowered both systolic and diastolic blood pressure; the diets with polyphenol-free olive oil did not alter blood pressure levels. In addition, the improvements were greater in the women who had a blood pressure that was higher at baseline.[9]

A similar study, which was published in 2008 in the *European Journal of Clinical Nutrition*, was conducted in Spain on 28 subjects with stable coronary disease. During two separate three-week periods, the subjects had daily supplementation of either 50mL of virgin olive oil or 50mL of refined olive oil. The researchers found that the virgin olive oil supplement reduced Interleukin-6 and C-reactive protein, two markers of inflammation. The researchers concluded that the "consumption of virgin olive oil could provide beneficial effects in stable coronary heart disease patients as an additional intervention to the pharmacological treatment."[10]

In an earlier study, published in 2004 in *Clinical Nutrition*, researchers from Spain and Argentina wanted to learn if virgin olive oil would be useful for elderly people with high blood pressure. The cohort consisted of 42 women and 20 men who lived at a residential home for the elderly in Spain. Thirty-one of the participants had high blood pressure; 31 had normal blood pressure. During the trial, the participants consumed diets enriched with virgin olive oil or sunflower oil for four weeks each, with a four-week washout period between them. The researchers found that virgin olive oil improved the systolic blood pressure in the men and women with elevated blood pressure. This may enable them to "reduce the requirement of anti-hypertensive medication."[11]

In a study published in 2008 in the *Journal of Agricultural and Food Chemistry*, researchers from Spain noted that virgin olive oil contains the polyphenols hydroxytyrosol and hydroxytyrosol acetate. So, they decided to test the antiplatelet effect of these two polyphenols in rats and compare their results to acetylsalicylic acid (ASA). While ASA is known to support cardiovascular health by reducing the probability of platelet aggregation and thrombosis, it may create a number of problems such as gastrointestinal bleeding. All three compounds were administered orally for seven days. Although the researchers found that both polyphenols lowered platelet aggregation, hydroxytyrosol acetate was more effective than hydroxytyrosol. As expected, ASA also lowered platelet aggregation. The researchers concluded that their findings "may offer an explanation for the beneficial effects of VOO [virgin olive oil] in preventing cardiovascular events and open new perspectives toward the potential use of these polyphenols as an alternative to ASA in the prevention of arterial thrombotic events."[12]

In a study published in 2011 in *Clinical Nutrition*, researchers from Spain, Demark, Finland, Germany, and Italy, who have combined together to form the EurOlive study, recruited 200 healthy men and randomly assigned them to one of three intervention groups. For three weeks, the men consumed 25mL/day of olive oil that contained a high dose, a medium dose, or a low dose of phenolic compounds. The researchers learned that the daily doses of polyphenol-rich olive oil raised levels of an antibody that fights oxidized LDL. In so doing, the olive oil lowers the risk of the hardening of the arteries and heart disease.[13]

IS OLIVE OIL BENEFICIAL FOR CARDIOVASCULAR SUPPORT?

From the research reviewed, olive oil certainly appears to support cardiovascular health. Unless someone has an intolerance or allergy to olive oil, which is thought to be rare, extra-virgin olive oil should be a valuable addition to everyone's diet. This is particularly true for those at increased risk for cardiovascular problems.

NOTES

1. Ruiz-Canela, Miguel and Miguel A. Martínez-González. March 2011. "Olive Oil in the Primary Prevention of Cardiovascular Disease." *Maturitas* 68(3): 245–250.

2. Damasceno, N. R. T., A. Pérez-Heras, M. Serra et al. June 2011. "Crossover Study of Diets Enriched with Virgin Olive Oil, Walnuts or Almonds. Effects on Lipids and Other Cardiovascular Risk Markers." *Nutrition, Metabolism & Cardiovascular Diseases* 21(Supplement 1): S14–S20.

3. Samieri, C., C. Féart, C. Proust-Lima et al. August 2, 2011. "Olive Oil Consumption, Plasma Oleic Acid, and Stroke Incidence." *Neurology* 77(5): 418–425.

4. Bendinelli, Benedetta, Giovanna Masala, Calogero Saieva et al. February 2011. "Fruit, Vegetables, and Olive Oil and Risk of Coronary Heart Disease in Italian Women: The EPICOR Study." *The American Journal of Clinical Nutrition* 93(2): 275–283.

5. Soriguer, F., G. Rojo-Martínez, A. Goday et al. 2013. "Olive Oil Has a Beneficial Effect on Impaired Glucose Regulation and Other Cardiometabolic Risk Factors." *European Journal of Clinical Nutrition* 67(9): 911–916. This is correct.

6. Widmer, R. J., M. A. Freund, A. J. Flammer et al. 2013. "Beneficial Effects of Polyphenol-Rich Olive Oil in Patients with Early Atherosclerosis." *European Journal of Nutrition* 52(3): 1223–1231.

7. Estruch, R., E. Ros, J. Salas-Salvadó et al. April 4, 2013. "Primary Prevention of Cardiovascular Disease with a Mediterranean Diet." *The New England Journal of Medicine* 368(14): 1279–1290.

8. Buckland, Genevieve, Noemic Travier, Aurelio Barricarte et al. 2012. "Olive Oil Intake and CHD in the European Prospective Investigation into Cancer and Nutrition Spanish Cohort." *British Journal of Nutrition* 108(11): 2075–2082.

9. Moreno-Luna, R., R. Muñoz-Hernandez, M. L. Miranda et al. December 2012. "Olive Oil Polyphenols Decrease Blood Pressure and Improve Endothelial Function in Young Women with Mild Hypertension." *American Journal of Hypertension* 25(12): 1299–1304.

10. Fitó, M., M. Cladellas, R. de la Torre et al. 2008. "Anti-Inflammatory Effect of Virgin Olive Oil in Stable Coronary Disease Patients: A Randomized, Crossover, Controlled Trial." *European Journal of Clinical Nutrition* 62(4): 570–574.

11. Perona, Javier S., Julio Cañizares, Emilio Montero et al. October 2004. "Virgin Olive Oil Reduces Blood Pressure in Hypertensive Elderly Subjects." *Clinical Nutrition* 23(5): 1113–1121.

12. González-Correa, José Antonio, María Dolores Navas, Javier Muñoz-Marín et al. 2008. "Effects of Hydroxytyrosol and Hydroxytyrosol Acetate Administration to Rats on Platelet Function Compared to Acetylsalicylic Acid." *Journal of Agricultural and Food Chemistry* 56(17): 7872–7876.

13. Castañer, Olga, Montserrat Fitó, M. Carmen López-Sabater et al. 2011. "The Effect of Olive Oil Polyphenols on Antibodies against Oxidized LDL. A Randomized Clinical Trial." *Clinical Nutrition* 30(4): 490–493.

REFERENCES AND RESOURCES

Magazines, Journals, and Newspapers

Bendinelli, Benedetta, Giovanna Masala, Calogero Saieva et al. February 2011. "Fruit, Vegetables, and Olive Oil and Risk of Coronary Heart Disease in Italian Women: The EPICOR Study." *The American Journal of Clinical Nutrition* 93(2): 275–283.

Buckland, Genevieve, Noemie Travier, Aurelio Barricarte et al. 2012. "Olive Oil Intake and CHD in the European Prospective Investigation into Cancer and Nutrition Spanish Cohort." *British Journal of Nutrition* 108(11): 2075–2082.

Castañer, Olga, Montserrat Fitó, M. Carmen López-Sabater et al. August 2011. "The Effect of Olive Oil Polyphenols on Antibodies against Oxidized LDL: A Randomized Clinical Trial." *Clinical Nutrition* 30(4): 490–493.

Damasceno, N. R. T., A. Pérez-Heras, M. Serra et al. June 2011. "Crossover Study of Diets Enriched with Virgin Olive Oil, Walnuts or Almonds. Effects on Lipids and Other Cardiovascular Risk Markers." *Nutrition, Metabolism & Cardiovascular Diseases* 21(Supplement 1): S14–S20.

Estruch, R., E. Ros, J. Salas-Salvadó et al. April 4, 2013. "Primary Prevention of Cardiovascular Disease with a Mediterranean Diet." *The New England Journal of Medicine* 368(14): 1279–1290.

Fitó, M., M. Cladellas, R. de la Torre et al. 2008. "Anti-Inflammatory Effect of Virgin Olive Oil in Stable Coronary Disease Patients: A Randomized, Crossover, Controlled Trial." *European Journal of Clinical Nutrition* 62(4): 570–574.

González-Correa, José Antonio, Mariá Dolores Navas, Javier Muñoz-Marín et al. 2008. "Effects of Hydroxytyrosol and Hydroxytyrosol Acetate Administration to Rats on Platelet Function Compared to Acetylsalicylic Acid." *Journal of Agricultural and Food Chemistry* 56(17): 7872–7876.

Lockwood, Brian and Emily Waterman. December 2007. "Active Components and Clinical Applications of Olive Oil." *Alternative Medicine Review* 12(4): 331–342.

Moreno-Luna, R., R. Muñoz-Hernandez, M. L. Miranda et al. December 2012. "Olive Oil Polyphenols Decrease Blood Pressure and Improve Endothelial Function in Young Women with Mild Hypertension." *American Journal of Hypertension* 25(12): 1299–1304.

Perona, Javier S., Julio Cañizares, Emilio Montero et al. October 2004. "Virgin Olive Oil Reduces Blood Pressure in Hypertensive Elderly Subjects." *Clinical Nutrition* 23(5): 1113–1121.

Ross, Stephanie Maxine. September–October 2013. "Effects of Extra Virgin Olive Oil Phenolic Compounds and the Mediterranean Diet on Cardiovascular Health." *Holistic Nursing Practice* 27(5): 303–307.

Ruiz-Canela, Miguel and Miguel A. Martínez-González. March 2011. "Olive Oil in the Primary Prevention of Cardiovascular Disease." *Maturitas* 68(3): 245–250.

Samieri, C., C. Féart, C. Proust-Lima et al. August 2, 2011. "Olive Oil Consumption, Plasma Oleic Acid, and Stroke Incidence." *Neurology* 77(5): 418–425.

Soriguer, F., G. Rojo-Martínez, A. Goday et al. 2013. "Olive Oil Has a Beneficial Effects on Impaired Glucose Regulation and Other Cardiometabolic Risk Factors." *European Journal of Clinical Nutrition* 67(9): 911–916.

Widmer, R. J., M. A. Freund, A. J. Flammer et al. 2013. "Beneficial Effects of Polyphenol-Rich Olive Oil in Patients with Early Atherosclerosis." *European Journal of Nutrition* 52(3): 1223–1231.

Websites

The George Mateljan Foundation. www.whfoods.com.
The Olive Oil Times. www.oliveoiltimes.com.

Olive Oil and Diabetes

This entry explores olive oil's usefulness in combating type 2 diabetes.

MAY BE USEFUL FOR PEOPLE WITH TYPE 2 DIABETES OR METABOLIC SYNDROME

In a study published in 2014 in *Free Radical Biology and Medicine*, researchers from China wanted to learn if hydroxytyrosol, which is a major polyphenolic compound in virgin olive oil, would help protect against metabolic syndrome. (Also known as insulin resistance, metabolic syndrome is a disorder in which a person had excess weight, high blood pressure, and elevated levels of cholesterol. It may lead to the development of type 2 diabetes.) During one part of the study, the researchers divided their mice into four groups. The mice in the first group were fed a normal diet, the mice in the second group were fed a high-fat diet, the mice in the third group were fed a high-fat diet that was supplemented with low doses of hydroxytyrosol, and the mice in the fourth group were fed a high-fat diet that was supplemented with high doses of hydroxytyrosol. After 17 weeks, the mice were sacrificed. The researchers found that supplementation with the higher dose of hydroxytyrosol inhibited the weight gain associated with the high-fat diet. Both doses of hydroxytyrosol lowered levels of fasting glucose and insulin. The researchers concluded that hydroxytyrosol had the potential to "normalize obesity, diabetes, dyslipidemia, inflammation, fatty liver, and insulin resistance induced by HFD [high-fat diet] feeding in rats."[1]

In a population-based, cross-sectional, cluster sampling study published in 2013 in the *European Journal of Clinical Nutrition*, researchers from multiple locations in Spain examined the association between consumption of olive oil and a number of medical problems including impaired glucose tolerance. The cohort consisted of 4,572 people over the age of 18. Information was obtained from an interviewer-administered structured questionnaire that was followed by a physical examination from a nurse and laboratory testing. The researchers determined that the people who consumed olive oil were less likely than those who consumed sunflower oil to have impaired glucose regulation. Moreover, "the greater protection against insulin resistance from olive oil occurred in those persons with obesity or a carbohydrate metabolism disorder."[2]

In a cross-sectional study published in 2011 in *Clinical Nutrition*, researchers from Spain investigated the association between the consumption of oleic acid, found in olive oil, and insulin resistance in people with elevated lipid levels. The cohort consisted of 205 nondiabetic men and 156 nondiabetic women, with mean ages of 44 and 46, respectively. The researchers found that the men and women with higher levels of oleic acid in their blood were less likely to have insulin resistance. Still, the researchers noted that "the extent to which this association is only attributable to OA [oleic acid] deserves further research, given the dietary polyphenols (antioxidant compounds naturally present in virgin olive oil) are suggested to modulate insulin sensitivity."[3]

In a cross-sectional study published in 2009 in *Nutrition, Metabolism & Cardiovascular Disease*, researchers from Spain evaluated the association between adherence to a Mediterranean diet and the risk for developing metabolic syndrome. The cohort consisted of 808 participants who were considered to be at an increased risk for cardiovascular disease. Though metabolic syndrome occurred more often in the women than the men, it was only in men that the researchers observed a significant inverse association between the adherence to a Mediterranean diet and incidence of metabolic syndrome. "Men at the top quartile of adherence to the MedDiet [Mediterranean diet] had a statistically significant lower odds ratio of having MetS [metabolic syndrome] in comparison with the lowest quartile." While the association was similar for women, their results were not statistically significant. The researchers concluded that their main finding was that the "highest adherence" to a Mediterranean diet was association with a "lower prevalence" of metabolic syndrome.[4]

In a study published in 2011 in *Diabetes Care*, researchers from Spain compared the rates in which the participants developed diabetes when they were on one of two Mediterranean-style diets—one with 1L/week of extra-virgin olive oil and the other with 30g/day of mixed nuts—or a low-fat diet. The cohort consisted of 418 men, between the ages of 55 and 80, and women, between the ages of 60 and 80. At baseline, none of the participants had cardiovascular disease, but they all had an increased risk because of medical problems such as high blood pressure, elevated levels of cholesterol, family history of premature cardiovascular disease, or a history of smoking. After a median follow-up of four years, 54 people developed new-onset diabetes. The incidence of diabetes was 10.1% in the olive oil group, 11.0% in the nut group, and 17.9% in the low-fat group. Moreover, "when the two MedDiet [Mediterranean Diet] groups were merged into a single category . . ., diabetes incidence was reduced by 52%."[5]

In a study published in 2013 in *Metabolic Syndrome and Related Disorders*, researchers from Italy investigated the ability of the Mediterranean diet to attenuate the symptoms associated with metabolic syndrome. The cohort consisted of 120 subjects (64.2% women) who were free of diabetes and cardiovascular disease at baseline, but were considered at increased risk for cardiovascular problems. All of the participants, who had a mean age of 59.8, had face-to-face interviews and a number of different medical tests. The researchers determined that the 44 subjects who were found to have metabolic syndrome were less likely to consume olive oil as "the main culinary fat." The researchers noted that their findings "support the hypothesis that dietary habits in line with the traditional Med Diet pattern may beneficially impact metabolic syndrome, prediabetes, and microinflammation in subjects free of diabetes and clinical cardiovascular disease."[6]

In a cross-sectional study published in 2010 in the *Journal of the American College of Nutrition*, researchers from Greece wanted to learn the association between the diets consumed by overweight and obese people from Greece and the prevalence of metabolic syndrome. The cohort consisted of 169 women and 30 men who were all overweight or obese. Initially, the participants were divided into two groups—those who had a Mediterranean-style diet and those who did not. The people who did not eat a Mediterranean diet were then divided into two groups—those who ate a high-carbohydrate diet and those who ate a high-fat diet. The vast majority of the participants, 154 people, ate a high-fat diet. Thirty-four ate a high-carbohydrate diet and 38 ate a Mediterranean-style diet. The researchers found that the Mediterranean-style diet was associated with "a decreased prevalence of metabolic syndrome." And, they concluded that "adherence to the MD [Mediterranean diet] could lead to a significant decrease in the prevalence of the metabolic syndrome and associated cardiovascular risk."[7]

In a meta-analysis published in 2011 in the *Journal of the American College of Cardiology*, researchers from Greece and Italy evaluated the effects of a Mediterranean-style diet on metabolic syndrome and its components. Their examination included 50 studies with a total of 534,906 participants. The researchers found that the Mediterranean diet was related to a lower incidence of metabolic syndrome. "Adherence to the Mediterranean dietary pattern was associated with lower MS [metabolic syndrome] prevalence and progression." In addition, the researchers determined that the "greater adherence to this traditional dietary pattern was associated with favorable effects on the MS components."[8]

IS OLIVE OIL BENEFICIAL FOR PEOPLE WITH TYPE 2 DIABETES AND METABOLIC SYNDROME?

From the research reviewed, there appears to be an abundance of evidence that people who have type 2 diabetes or metabolic syndrome should include olive oil in their diets. And, there is very good evidence that olive oil may well help improve some of the symptoms associated with these medical problems.

NOTES

1. Cao, Ke, Jie Xu, Xuan Zou et al. 2014. "Hydroxytyrosol Prevents Diet-Induced Metabolic Syndrome and Attenuates Mitochondrial Abnormalities in Obese Mice." *Free Radical Biology and Medicine* 67: 396–407.

2. Soriguer, F., G. Rojo-Martínez, A. Goday et al. September 2013. "Olive Oil Has a Beneficial Effect on Impaired Glucose Regulation and Other Cardiometabolic Risk Factors. Dia@bet.es Study." *European Journal of Clinical Nutrition* 67(9): 911–916.

3. Sala-Vila, A., M. Cofán, R. Mateo-Gallego et al. 2011. "Inverse Association between Serum Phospholipid Oleic Acid and Insulin Resistance in Subjects with Primary Dyslipidaemia." *Clinical Nutrition* 30(5): 590–592.

4. Babio, N., M. Bulló, J. Basora et al. 2009. "Adherence to the Mediterranean Diet and Risk of Metabolic Syndrome and Its Components." *Nutrition, Metabolism & Cardiovascular Diseases* 19(8): 563–570.

5. Salas-Salvadó, Jordi, Monica Bulló, Nancy Babio et al. January 2011. "Reduction in the Incidence of Type 2 Diabetes with the Mediterranean Diet Results of the PREDIMED-Reus Nutrition Intervention Randomized Trial." *Diabetes Care* 34(1): 14–19.

6. Viscogliosi, Giovanni, Elisa Cipriani, Maria Livia Liguori et al. 2013. "Mediterranean Dietary Pattern Adherence: Associations with Prediabetes, Metabolic Syndrome, and Related Microinflammation." *Metabolic Syndrome and Related Disorders* 11(3): 210–216.

7. Paletas, Konstantinos, Eleni Athanasiadou, Maria Sarigianni et al. 2010. "The Protective Rose of the Mediterranean Diet on the Prevalence of Metabolic Syndrome in a Population of Greek Obese Subjects." *Journal of the American College of Nutrition* 29(1): 41–45.

8. Kastorini, Christina-Maria, Haralampos J. Milionis, Katherine Esposito et al. 2011. "The Effect of Mediterranean Diet on Metabolic Syndrome and Its Components." *Journal of the American College of Cardiology* 57(11): 1299–1313.

REFERENCES AND RESOURCES

Magazines, Journals, and Newspapers

Babio, N., M. Bulló, J. Basora et al. 2009. "Adherence to the Mediterranean Diet and Risk of Metabolic Syndrome and Its Components." *Nutrition, Metabolism & Cardiovascular Diseases* 19(8): 563–570.

Cao, Ke, Jie Xu, Xuan Zou et al. 2014. "Hydroxytyrosol Prevents Diet-Induced Metabolic Syndrome and Attenuates Mitochondrial Abnormalities in Obese Mice." *Free Radical Biology and Medicine* 67: 396–407.

Kastorini, Christina-Maria, Haralampos J. Milionis, Katherine Esposito et al. 2011. "The Effect of Mediterranean Diet on Metabolic Syndrome and Its Components." *Journal of the American College of Cardiology* 57(11): 1299–1313.

Paletas, Konstantinos, Eleni Athanasiadou, Maria Sarigianni et al. 2010. "The Protective Role of the Mediterranean Diet on the Prevalence of Metabolic Syndrome in a Population of Greek Obese Subjects." *Journal of the American College of Nutrition* 29(1): 41–45.

Sala-Vila, A., M. Cofán, R. Mateo-Gallego et al. 2011. "Inverse Association between Serum Phospholipid Oleic Acid and Insulin Resistance in Subjects with Primary Dyslipidaemia." *Clinical Nutrition* 30(5): 590–592.

Salas-Salvadó, Jordi, Monica Bulló, Nancy Babio et al. January 2011. "Reduction in the Incidence of Type 2 Diabetes with the Mediterranean Diet Results of the PREDIMED-Reus Nutrition Intervention Randomized Trial." *Diabetes Care* 34(1): 14–19.

Soriguer, F., G. Rojo-Martínez, A. Goday et al. September 2013. "Olive Oil Has a Beneficial Effect on Impaired Glucose Regulation and Other Cardiometabolic Risk Factors. Dia@bet.es Study." *European Journal of Clinical Nutrition* 67(9): 911–916.

Viscogliosi, Giovanni, Elisa Cipriani, Maria Livia Liguori et al. June 2013. "Mediterranean Dietary Pattern Adherence: Association with Prediabetes, Metabolic Syndrome, and Related Microinflammation." *Metabolic Syndrome and Related Disorders* 11(3): 210–216.

Website

International Olive Council. www.interantionaloliveoil.org.

Olive Oil and Memory Function

This entry discusses olive oil's potential for supporting healthy memory function.

SUPPORTS MEMORY

In a study published in 2012 in *Rejuvenation Research*, researchers from Italy evaluated the ability of olive oil phenols to improve the memory of aging mice. The mice were randomly divided into three groups of 20 mice. The mice in each group were fed different diets for 12 months. The diet in one group of mice was supplemented with extra-virgin olive oil high in phenolic antioxidants. A second group of mice received the same extra-virgin olive oil without the phenolic antioxidants. And, the third group of mice took supplemental extra-virgin olive oil and resveratrol, a natural phenol with antiaging properties. During the study, the researchers conducted a number of tests on the mice, including tests that assessed memory. At the end of the study, the mice were sacrificed, and their brains were evaluated. The researchers found that the administration of olive oil supplementation reversed some of the problems associated with aging without causing any harm. Mice fed the olive oil with high amount of phenols did very well in a number of tests including the memory testing, and they did better than the mice fed the olive oil mixed with resveratrol. The researchers concluded that "natural polyphenols contained in extra-virgin olive oil can improve some age-related dysfunctions by differentially affecting different brain areas." Moreover, "such a modulation can be obtained with an olive oil intake that is normal in the Mediterranean area, provided that the oil has a sufficiently high content of polyphenols."[1]

In a study published in 2013 in *The Journal of Nutrition, Health & Aging*, researchers from Spain examined the association between virgin olive oil supplementation and long-term cognition. The 271 subjects were randomly placed on one of three different diets—a Mediterranean diet supplemented with extra-virgin olive oil, a Mediterranean diet supplemented with nuts, or a low-fat diet. Two hundred and sixty-eight subjects completed the 6.5-year study. The researchers found that the subjects in the extra-virgin olive oil supplementation group had "better cognition especially across fluency and memory tasks and less MCI [mild cognitive impairment] as compared to the controls." In addition, when compared to the subjects taking nuts, the subjects on extra-virgin olive oil "had a significantly better performance on both visual and verbal memory domains."[2]

In a study published in 2013 in *ACS Chemical Neuroscience*, researchers from Louisiana (United States) investigated the ability of oleocanthal, a phenolic component of extra-virgin olive oil, to reduce the risk of Alzheimer's disease, an illness characterized by the accumulation of beta-amyloid and tau proteins in the brain. Among the many problems experienced by people with Alzheimer's disease is the profound loss of memory. In their studies of mice and the cultured brain cells of mice, the researcher traced the effects of oleocanthal. They found that oleocanthal increased the production of proteins and enzymes that remove beta-amyloid from the brain. The researchers concluded that "extra-virgin olive oil-derived oleocanthal associated with the consumption of Mediterranean diet has the potential to reduce the risk of AD [Alzheimer's disease] or related neuro-degenerative dementias."[3]

In a study published in 2009 in *Dementia and Geriatric Cognitive Disorders*, researchers from France and the United Kingdom examined the association between the use of olive oil and cognitive deficits and cognitive decline in an elderly, independently living population of 6,947 subjects. According to their use of olive oil, the subjects were placed in one of three groups. Those who used olive oil for cooking and dressing were considered intensive users, those who used olive oil for cooking or dressing were moderate users, and there were also subjects who used no olive oil. Cognitive tests were administered during a four-year follow-up period. The researchers learned that the subjects who were intensive or moderate uses of olive oil were less likely than those who never used olive oil to have cognitive deficits for verbal fluency and visual memory. According to the researchers, their study demonstrated that "in a large non-demented elderly population . . . olive oil consumption habits are significantly associated with selective cognitive deficit and cognitive decline, independent of other dietary intakes and after adjusting for potential confounders."[4]

In a study published in 2013 in *Neurology*, researchers from Greece and several locations in the United States wanted to learn if the consumption of a Mediterranean diet, including olive oil, would help older people retain more of their memory. The cohort consisted of 17,478 men and women with a mean age of 64.4. Of these, 9,181 of the participants (53%) had a low adherence to a Mediterranean diet. During the four years of the study, the participants participated in tests that measured their memory and cognitive skills. In general, the researchers found that those who followed the Mediterranean diet were 19% less likely to develop memory and cognitive problems. However, there was an exception. For the 17% of the participants who had diabetes, the Mediterranean diet did not appear to prevent memory and cognitive problems from developing. The researchers concluded that the "higher adherence to MeD [Mediterranean diet] was associated with a lower likelihood of ICI [incident cognitive impairment] independent of potential cofounders. The association was moderated by presence of diabetes mellitus."[5]

In a study published in 2009 in the *Archives of Neurology*, researchers from New York City investigated the association between consumption of a Mediterranean diet, including olive oil, and the incidence of mild cognitive impairment and the progression to Alzheimer's disease. The initial cohort consisted of 1,393 subjects without cognitive impairment and 482 people with mild cognitive impairment. After a follow-up period that averaged 4.5 years, 275 of the subjects who did not have cognitive impairment developed the condition. When compared to the one-third of the subjects who had the lowest scores for consumption of a Mediterranean diet, the one-third of the subjects with the highest scores had a 28% lower risk of development mild cognitive impairment. The subjects in the middle one-third group had a 17% lower risk than those in the lowest Mediterranean diet consumption group. Meanwhile, of the 482 subjects with mild cognitive impairment at the beginning of the study, 106 went on to develop Alzheimer's disease during an average of 4.3 years of follow-up. Again, adhering to a Mediterranean diet reduced the risk of Alzheimer's disease. When compared to the one-third of the subjects with the lowest intake of a Mediterranean diet, the one-third of the subjects with the highest scores for a Mediterranean diet had 48% less risk and those in the middle third had a 45% reduced risk. The researchers concluded that a higher adherence to the Mediterranean diet "is associated with a borderline reduction in the risk of developing MCI [mild cognitive impairment] and a reduction in the risk of conversion from MCI to AD [Alzheimer's disease]."[6]

In a study published in 2012 in the *Journal of Alzheimer's Disease*, researchers based in St. Louis, Missouri (United States), and also from several other locations, divided mice into four groups. For four weeks, the mice in one group received supplemental extra-virgin olive oil, the mice in another group received supplemental extra-virgin olive oil with extra polyphenols, the mice in a third group received supplemental coconut oil, and the mice in a fourth group received supplemental butter. While on the supplementation, the mice were given a variety of different tests to evaluate memory and cognition. The researchers found that the mice taking extra-virgin olive oil with extra polyphenols had the best results. The mice in the second best group were those on extra-virgin olive oil. The mice taking coconut oil saw improvements in memory, but their learning ability was about the same as the mice fed butter. According to the researchers, their findings "suggest that EVOO [extra-virgin olive oil] has beneficial effects on learning and memory deficits found in aging and diseases, such as those related to the overproduction of amyloid-β protein, by reversing oxidative damage in the brain, effects that are augmented with increasing concentrations of polyphenols in EVOO."[7]

IS OLIVE OIL BENEFICIAL FOR MEMORY AND COGNITIVE?

From the research reviewed, olive oil does appear to improve memory and cognition. This is especially important information for the millions of aging people throughout the world.

NOTES

1. Pitozzi, Vanessa, Michela Jacomelli, Dolores Catelan et al. December 2012. "Long-Term Dietary Extra-Virgin Olive Oil Rich in Polyphenols Reverses Age-Related Dysfunction in Motor Coordination and Contextual Memory in Mice: Role of Oxidative Stress." *Rejuvenation Research* 15(6): 601–612.

2. Martínez-Lapiscina, E.H., P. Clavero, E. Toledo et al. 2013. "Virgin Olive Oil Supplementation and Long-Term Cognition: The PREDIMED-NAVARRA Randomized Trial." *The Journal of Nutrition, Health & Aging* 17(6): 544–552.

3. Abuznait, Alaa H., Hisham Qosa, Belnaser A. Busnena et al. 2013. "Olive-Oil-Derived Oleocanthal Enhances β-Amyloid Clearance as a Potential Neuroprotective Mechanism against Alzheimer's Disease: In Vitro and In Vivo Studies." *ACS Chemical Neuroscience* 4(6): 973–982.

4. Berr, Claudine, Florence Portet, Isabelle Carriere et al. 2009. "Olive Oil and Cognition: Results from the Three-City Study." *Dementia and Geriatric Cognitive Disorders* 28(4): 357–364.

5. Tsivgoulis, Georgios, Suzanne Judd, Abraham J. Letter et al. April 30, 2013. "Adherence to a Mediterranean Diet and Risk of Incident Cognitive Impairment." *Neurology* 80(18): 1684–1692.

6. Scarmeas, Nikolaos, Yaakove Stern, Richard Mayeux et al. February 2009. "Mediterranean Diet and Mild Cognitive Impairment." *Archives of Neurology* 66(2): 216–225.

7. Farr, S. A., T. O. Price, L. J. Dominguez et al. 2012. "Extra Virgin Olive Oil Improves Leaning and Memory in SAMP8 Mice." *Journal of Alzheimer's Disease* 28(1): 81–92.

REFERENCES AND RESOURCES

Magazines, Journals, and Newspapers

Abuznait, Alaa H., Hisham Qosa, Belnaser A. Busnena et al. June 19, 2013. "Olive-Oil-Derived Oleocanthal Enhances β-Amyloid Clearance as a Potential Neuroprotective Mechanism against Alzheimer's Disease: In Vitro and In Vivo Studies." *ACS Chemical Neuroscience* 4(6): 973–982.

Berr, Claudine, Florence Portet, Isabelle Carriere et al. October 30, 2009. "Olive Oil and Cognition: Results from the Three-City Study." *Dementia and Geriatric Cognitive Disorders* 28(4): 357–364.

Farr, S. A., T. O. Price, L. J. Dominguez et al. 2012. "Extra Virgin Olive Oil Improves Learning and Memory in SAMP8 Mice." *Journal of Alzheimer's Disease* 28(1): 81–92.

Martínez-Lapiscina, E. H., P. Clavero, E. Toledo et al. 2013. "Virgin Olive Oil Supplementation and Long-Term Cognition: The PREDIMED-NAVARRA Randomized Trial." *The Journal of Nutrition, Health & Aging* 17(6): 544–552.

Pitozzi, Vanessa, Michela Jacomelli, Dolores Catelan et al. December 2012. "Long-Term Dietary Extra-Virgin Olive Oil Rich in Polyphenols Reverses Age-Related Dysfunctions in Motor Coordination and Contextual Memory in Mice: Role of Oxidative Stress." *Rejuvenation Research* 15(6): 601–612.

Scarmeas, Nikolaos, Yaakove Stern, Richard Mayeux et al. February 2009. "Mediterranean Diet and Mild Cognitive Impairment." *Archives of Neurology* 66(2): 216–225.

Tsivgoulis, Georgios, Suzanne Judd, Abraham J. Letter et al. April 30, 2013. "Adherence to a Mediterranean Diet and Risk of Incident Cognitive Impairment." *Neurology* 80(18): 1684–1692.

Website

Alzheimer's Association. Alz.org.

Olive Oil and Skin

Olive oil has long been recognized as supporting healthy skin.

SUPPORTS SKIN HEALTH

In a cross sectional study published in 2012 in *PLoS One,* researchers from France and Austria investigated the association between intake of monounsaturated fatty acids, such as olive oil, and the risk of skin "photoaging," or damage to the skin from ultraviolet radiation. The cohort consisted of 1,264 women and 1,655 men between the ages of 45 and 60. The subjects were asked to complete a 24-hour dietary record every two months for a total of six records per year. "Ten records were considered to be sufficient to estimate the individual intake of monounsaturated fatty acids with acceptable accuracy." Meanwhile, trained investigators assessed the severity of photoaging.

The researchers found that in both men and women, there was a significant association between the severity of photoaging and the dietary intake of monounsaturated fatty acids. People who consumed higher amounts of monounsaturated fatty acids had a reduced risk of severe photoaging. This was most evident with olive oil, the primary source of monounsaturated fats in this cohort. "A higher intake of olive oil was significantly associated with a lower risk of severe photoaging."[1]

In a study published in 2010 in *Pharmacognosy Research,* researchers from India evaluated the sun protection values of herbal oils used in sunscreens and cosmetics. "Sun protection factor is a laboratory measure of the effectiveness of sunscreen; the higher the SPF, the more protection a sunscreen offers against the ultraviolet radiation causing sunburn." Of the many oils that the researchers tested, olive oil and coconut oil proved to provide the most sun protection. According to the researchers, information on the sun protection values is useful for people deciding what to include in their cosmetic formulations. "Oil is the most important constituent of creams and lotions."[2]

In a study published in 2013 in the *African Journal of Traditional, Complementary and Alternative Medicine,* researchers from Turkey wanted to investigate the cytotoxic and antibacterial properties of olive oil and lime cream (made from limestone), both individually and combined. They tested these properties in a cell culture on six microbial strains and two yeast strains. The researchers learned that lime cream had "antibacterial activity but also cytotoxic on the fibroblasts." Meanwhile, olive oil had "limited or no antibacterial effect and it had little or no cytotoxic on the fibroblasts." When they combined olive oil with lime cream, the researchers found that "olive oil reduced its cytotoxic impact."[3]

In a study published in 2009 in the *Journal of Korean Medical Science,* researchers from Korea compared the ability of pure olive oil, ozonated olive oil, and no treatment to heal skin wounds on the backs of 16 guinea pigs. The researchers began by creating four wounds on the backs of each guinea pig. Every day, the researchers applied pure olive oil to one wound and ozonated olive oil to two wounds. In order to serve as a control, the fourth wound was not treated. On day 3, four pigs were euthanized; on day 7, another four were euthanized; and the final pigs were put down on day 11. The researchers found that the ozonated olive oil "significantly enhanced the acute cutaneous wound healing." In addition, "the ozone group showed a significantly decreased residual wound area as compared to the oil group, as well as the control group."[4]

In a study published in 2006 in *Phytomedicine*, researchers based in Italy conducted laboratory studies to examine the antifungal properties of olive tree fruit. They noted that it had already been determined that the topical application of olive oil mixed with honey and beeswax was useful in the treatment of fungal infections and diaper dermatitis. So, they studied the use of components in olive fruit against fungi. And, they found that these components had significant antifungal properties. As a result, adding olive oil components should enhance the effectiveness of skin disease preparations. In the final paragraph of their article, the researchers emphasized an important aspect of their findings. They noted that there are only a small number of substances that "are known to inhibit human pathogenic fungi, which are often completely resistant to antibiotics, and most of them are relatively toxic." Therefore, there has been an increase in efforts to find new antifungal agents, including those from plants. "These natural compounds could be useful agents in the topical treatment of fungal infections."[5]

In a trial published in 2012 in *Cutaneous and Ocular Toxicology*, researchers from Iran compared the ability of a cream made from olive oil and aloe vera to betamethasone 0.1% cream to treat chronic skin lesions caused by exposure to sulfur mustard (also known as mustard gas). The cohort consisted of 67 Iranians who were injured in war. For six weeks, the veterans were randomly assigned to apply one of the creams twice each day. The researchers found that the veterans in both groups had significant reductions in itching, burning, scaling, and dry skin. The researchers concluded that the olive oil and aloe vera cream "was at least as effective as betamethasone 0.1% in the treatment of sulfur mustard-induced chronic skin complications and might serve as a promising therapeutic option for the alleviation of symptoms in mustard gas-exposed patients."[6]

In a study published in 2008 in *Pediatric Dermatology*, researchers from Austria randomly assigned 173 preterm infants to receive one of three twice-daily topical treatments. One group of babies was treated with a water-in-oil emollient cream, a second group of babies was treated with an olive oil cream, and the third group of babies was treated with routine skin care (this group served as the control). Each baby was treated for a maximum of four weeks. The researchers found that the "neonates treated with olive oil cream showed statistically less dermatitis than did neonates treated with emollient cream, and both had a better outcome than those in the control group." Why is this finding so important?

"The skin of the preterm infant is an ineffective epidermal barrier." Such babies may lose large amount of water from their skin. An effective topical treatment may "improve skin integrity."[7]

In a study published in 2012 in the *Scandinavian Journal of Public Health*, researchers from Italy investigated the effect of an adherence to a Mediterranean diet and the incidence of acne. The cohort consisted of 93 people (median age of 17) with mild, moderate, and severe acne and 200 controls (median age of 16). The researchers found that the healthy controls followed a diet that was "closer to the Mediterranean diet" than the diet of those people who had acne. According to the researchers, "this is the first study demonstrating a protective role of the Mediterranean diet in the pathogenesis of acne."[8]

But, the research literature on olive oil and the skin does contain isolated instances of problems. In an article published in 2008 in *Contact Dermatitis*, a case report from the Netherlands described a 66-year-old man who was treated for cheilitis (painful inflammation and cracking of the corners of the mouth) with a product that contained 70% olive oil. Although it is believed to be rare, he developed a contact dermatitis to the olive oil. He was told to discontinue the treatment and not to consume olive oil in foods. The authors of the report concluded that "the rare possibility of sensitization to olive oil must be taken into consideration especially if it is used as a topical ointment, especially under occlusion."[9]

Two years earlier, also in *Contact Dermatitis*, a case report from Australia described a 40-year-old female with a five-month history of a rash on the thighs, arms, hands, and face. While being interviewed, the woman reported that she had been working as an aromatherapist for a year and that part of her job entailed mixing various essential oils with olive oil, a carrier oil. These oils were generally massaged into the skin of her clients. Interesting, when she was not working her condition improved. After testing, her medical provider suggested she stop using olive oil. "The patient was able to continue as an aromatherapist using soybean oil as the carrier oil."[10]

IS OLIVE OIL BENEFICIAL FOR SKIN CONDITIONS?

From the research reports presented in this entry, olive oil appears to be useful for a number of different skin conditions. And, apparently, allergic reactions to olive oil are quite rare.

NOTES

1. Latreille, Julie, Emmanuelle Kesse-Guyot, Denis Malvy et al. 2012. "Dietary Monounsaturated Fatty Acids Intake and Risk of PhotoAging." *PLoS One* 7(9): e44490.

2. Kaur, C. D. and S. Saraf. January 2010. "*In Vitro* Sun Protection Factor Determination of Herbal Oils Used in Cosmetics." *Pharmacognosy Research* 2(1): 22–25.

3. Sumer, Zeynep, Gulay Yildirim, Haldun Sumer, and Sahin Yildirim. May 16, 2013. "Cytotoxic and Antibacterial Activity of the Mixture of Olive Oil and Lime Cream *In*

Vitro Conditions." *African Journal of Traditional, Complementary and Alternative Medicine* 10(4): 137–143.

4. Kim, H.S., S.U. Noh, Y.W. Han et al. June 2009. "Therapeutic Effects of Topical Application of Ozone on Acute Cutaneous Wound Healing." *Journal of Korean Medical Science* 24(3): 368–374.

5. Battinelli, L., C. Daniele, M. Cristiani et al. September 2006. "*In Vitro* Antifungal and Anti-Elastase Activity of Some Aliphatic from *Olea europaea* L. Fruit." *Phytomedicine* 13(8): 558–563.

6. Panahi, Y., S.M. Davoudi, A. Sahebkar et al. June 2012. "Efficacy of Aloe Vera/ Olive Oil Cream versus Betamethasone Cream for Chronic Skin Lesions Following Sulfur Mustard Exposure: A Randomized Double-Blind Clinical Trial." *Cutaneous and Ocular Toxicology* 31(2): 95–103.

7. Kiechl-Kohlendorfer, U., C. Berger, and R. Inzinger. March–April 2008. "The Effect of Daily Treatment with an Olive Oil/Lanolin Emollient on Skin Integrity in Preterm Infants: A Randomized Controlled Trial." Pediatric Dermatology 25(2): 174–178.

8. Skroza, N., E. Tolino, L. Semyonov et al. July 2012. "Mediterranean Diet and Familial Dysmetabolism as Factors Influencing the Development of Acne." *Scandinavian Journal of Public Health* 40(5): 466–474.

9. Beukers, S.M., T. Rustemeyer, and D.P. Bruynzeel. October 2008. "Cheilitis due to Olive Oil." *Contact Dermatitis* 59(4): 253–255.

10. Williams, J.D. and B.J. Tate. October 2006. "Occupational Allergic Contact Dermatitis from Olive Oil." *Contact Dermatitis* 55(4): 251–252.

REFERENCES AND RESOURCES

Magazines, Journals, and Newspapers

Battinelli, L., C. Daniele, M. Cristiani et al. September 1, 2006. "*In Vitro* Antifungal and Anti-Elastase Activity of Some Aliphatic Aldehydes from *Olea europaea* L. Fruit." *Phytomedicine* 13(8): 558–563.

Beukers, S.M., T. Rustemeyer, and D.P. Bruynzeel. October 2008. "Cheilitis due to Olive Oil." *Contact Dermatitis* 59(4): 253–255.

Kaur, C. D and S. Saraf. January 2010. "*In Vitro* Sun Protection Factor Determination of Herbal Oils Used in Cosmetics." *Pharmacognosy Research* 2(1): 22–25.

Kiechl-Kohlendorfer, U., C. Berger, and R. Inzinger. March–April 2008. "The Effect of Daily Treatment with an Olive Oil/Lanolin Emollient on Skin Integrity in Preterm Infants: A Randomized Controlled Trial." *Pediatric Dermatology* 25(2): 174–178.

Kim, H.S., S.U. Noh, Y.W. Han et al. June 2009. "Therapeutic Effects of Topical Application of Ozone on Acute Cutaneous Wound Healing." *Journal of Korean Medical Science* 24(3): 368–374.

Latreille, Julie, Emmanuelle Kesse-Guyot, Denis Malvy et al. 2012. "Dietary Monounsaturated Fatty Acids Intake and Risk of Skin Photoaging." *PLoS One* 7(9): e44490.

Panahi, Y., S.M. Davoudi, A. Sahebkar et al. June 2012. "Efficacy of Aloe Vera/Olive Oil Cream versus Betamethasone Cream for Chronic Skin Lesions Following Sulfur Mustard Exposure: A Randomized Double-Blind Clinical Trial." *Cutaneous and Ocular Toxicology* 31(2): 95–103.

Skroza, N., E. Tolino, L. Semyonov et al. July 2012. "Mediterranean Diet and Familial Dysmetabolism as Factors Influencing the Development of Acne." *Scandinavian Journal of Public Health* 40(5): 466–474.

Sumer, Zeynep, Gulay Yildirim, Haldun Sumer, and Sahin Yildirim. May 16, 2013. "Cytotoxic and Antibacterial Activity of the Mixture of Olive Oil and Lime Cream *In Vitro* Conditions." *African Journal of traditional, Complementary and Alternative Medicine* 10(4): 137–143.

Williams, J. D. and B. J. Tate. October 2006. "Occupational Allergic Contact Dermatitis from Olive Oil." *Contact Dermatitis* 55(4): 251–252.

Website

Wellness Today. www.wellnesstoday.com.

Olive Oil and Weight Management

This entry examines how olive oil may impact weight management.

USEFUL FOR CONTROLLING WEIGHT

In a study published in 2012 in *Obesity Facts*, researchers from Spain examined the association between the consumption of olive oil and body mass index (BMI) and the risk of obesity in adults from Spain. The cohort consisted of 6,352 men and women. Food consumption was determined by a food frequency questionnaire. Only a tiny 1.5% of the men and 1.2% of the women reported that they consumed no olive oil. The researchers found that when compared to a daily consumption of less than one tablespoons of olive oil, a daily olive oil consumption of two tablespoons was not associated with a higher BMI or increased risk for obesity. And, they noted that "other studies have reported similar findings."[1]

In a study published in 2010 in the *Journal of Women's Health*, researchers from Rhode Island (United States), investigated the ability of an olive oil–enriched diet to help women who had previously been diagnosed with invasive breast cancer control their weight. Why is weight control among these women so important? Excess weight increases the risk of a recurrence.

The initial cohort consisted of 44 overweight women. All of these were diagnosed with invasive breast cancer after the age of 50 and were within four years of the completion of their treatment. Each woman consumed two diets for eight weeks. During the first eight weeks, they consumed the lower-fat diet recommended by the National Cancer Institute or an olive oil–enriched diet. During the second eight weeks, they followed the alternate diet. Both diets contained 1,500 calories/day; the olive oil–enriched diet included at least 3 tablespoons/day of extra-virgin olive oil. Twenty-eight women completed both weight loss diets. While all the participants lost weight, those who first followed the enriched olive oil diet lost more weight. Twenty-two women were then asked to select one of the diets and remain on it for six months. The vast majority selected the olive oil–enriched diet, which they found easier to follow than the low-fat diet. Twenty women completed the entire study. The researchers commented that "an olive oil-enriched diet may be more efficacious for weight loss in breast cancer survivors than a standard lower-fat diet."[2]

In a study published in 2006 in *Lipids*, researchers from Spain examined the association between the consumption of olive oil and the probability and incidence of weight gain in a cohort of 7,368 male and female Spanish university graduates, who were followed for a median period of 28.5 months. The researchers found that those who consumed higher amounts of olive oil were less likely to gain weight. According to the researchers, "the risk of obesity is not increased in subjects who follow an olive oil-rich Mediterranean food pattern." And, they concluded that "the belief that olive oil consumption increases the risk of obesity in Mediterranean populations is not supported by any reliable evidence."[3]

In a study published in 2009 in the *European Journal of Clinical Nutrition*, researchers from Spain tested the association between consumption of olive oil, sunflower oil, or a mixture of the oils and the incidence of obesity over a six-year period. The cohort consisted of 613 people between the ages of 18 and 65. The researchers found that "the incidence of obesity was significantly higher in those who consumed sunflower oil than in those who consumed olive oil or a mixture." Moreover, "persons who consumed olive oil had a lower risk of being obese at follow-up, adjusted for age, sex, physical activity, smoking, instruction level, energy intake and BMI [body mass index] in the baseline study."[4]

In a study published in 2011 in *Metabolic Syndrome and Related Disorders*, researchers from Italy and Greece conducted a meta-analysis of randomized controlled trials on the effect of a Mediterranean diet, including olive oil, on body weight. Their meta-analysis included a total of 3,436 participants—1,848 assigned to a Mediterranean diet and 1,588 assigned to a control diet. The researchers found that the Mediterranean diet "may be a useful tool to reduce body weight." This is particularly true when "the Mediterranean diet is energy-restricted, associated with physical activity, and with a follow-up of more than 6 months."[5]

In a study conducted at the Technische Universität München in Germany and at the University of Vienna, researchers divided 120 people into five groups. For three months, everyone ate 500g of yogurt each day. In four of the groups, the yogurt was enhanced with lard or butter or canola oil or olive oil. The members of the remaining group ate unenhanced yogurt. Periodic blood tests showed that those eating the olive oil–enhanced yogurt experienced increases in serotonin levels, a hormone associated with satiety. Like the butter and plain yogurt group, the members of the olive oil group did not gain weight; at the same time, the members of the lard and canola oil groups gained weight.

During the second part of the study, the subjects were placed in one of two groups. While members of each group ate nonfat yogurt, one type of yogurt was infused with an extract that had an olive oil aroma. The subjects who ate the olive oil–aromatic yogurt ate the same amount of calories; the subjects eating the plain yogurt increased their caloric intake. And, those eating the plain yogurt had less serotonin in their blood.[6]

In a study published in 2011 in the *European Journal of Endocrinology*, researchers from Spain compared the risk of weight gain among children who consumed olive oil to children who consumed sunflower oil or a mixture of sunflower and olive oils. The cohort consisted of 18 girls and 74 boys between the ages of 13 and 166 months who had undergone inguinal hernia surgery. (An inguinal hernia is the protrusion of tissue in the groin area.) A survey of eating habits and physical activity was conducted for each child; samples of subcutaneous adipose tissue were also taken. One year after baseline, all the children had follow-up visits.

The researchers learned that the food that 71.1% of the children ate was prepared only with olive oil. At the same time, 28.9% of the children ate food prepared with sunflower oil or the oil mixture. The researchers found that the children who consumed only olive oil were less likely to have increases in their BMI. And, they concluded that "diets with monounsaturated fatty acid (MUFA)-rich olive oil could reduce the risk of obesity in children."[7]

In a study published in 2009 in the *European Journal of Clinical Nutrition*, researchers from Spain analyzed the influence of a three-year intervention of three types of diet—a Mediterranean diet with a high intake of virgin olive oil or a Mediterranean diet with a high intake of nuts or a conventional low-fat diet—on 187 subjects at increased risk for cardiovascular problems. The researchers found that the subjects eating the Mediterranean diet, especially the diet that included a high intake of virgin olive oil, lost weight. They noted that the weight loss was "associated with [the diet's] higher levels of plasma antioxidant capacity."[8]

In a study published in 2008 in *Cell Metabolism*, researchers from California and Italy infused oleic acid, which is found in olive oil, into the small intestines of rats. Once in the intestines, oleic acid converted to oleoylethanolamide (OEA), a lipid hormone associated with both satiety and reductions in the consumption of food. The researchers noted that OEA is a "key physiological signal that specifically links dietary fat ingestion to across-meal satiety." And, they suggested that "nutritional and pharmacological strategies aimed at magnifying this lipid-sensing mechanism, such as inhibitors of OEA degradation, might be useful in the treatment of obesity and other eating disorders."[9]

IS OLIVE OIL HELPFUL FOR CONTROLLING WEIGHT?

From the research reviewed, olive oil certainly appears to help control weight. It even seems to be useful in facilitating weight loss.

NOTES

1. Benítez-Arciniega, A. D., D. Gómez-Ulloa, A. Vila et al. 2012. "Olive Oil Consumption, BMI, and Risk of Obesity in Spanish Adults." *Obesity Facts* 5(1): 52–59.

2. Flynn, Mary M. and Steven E. Reinert. June 2010. "Comparing an Olive Oil-Enriched Diet to a Standard Lower-Fat Diet for Weight Loss in Breast Cancer Survivors: A Pilot Study." *Journal of Women's Health* 19(6): 1155–1161.

3. Bes-Rastrollo, M., A. Sánchez-Villegas, C. de la Fuente et al. 2006. "Olive Oil Consumption and Weight Change: The SUN Prospective Cohort Study." *Lipids* 41(3): 249–256.

4. Soriguer, F., M.C. Almaraz, M.S. Ruiz-de-Adana et al. November 2009. "Incidence of Obesity Is Lower in Persons Who Consume Olive Oil." *European Journal of Clinical Nutrition* 63(11): 1371–1374.

5. Esposito, Katherine, Christina-Maria Kastorini, Demosthenes B. Panagiotakos, and Dario Giugliano. February 2011. "Mediterranean Diet and Weight Loss: Meta-Analysis of Randomized Controlled Trials." *Metabolic Syndrome and Related Disorders* 9(1): 1–12.

6. Science Daly Website. http://www.sciencedaily.com/releases/2013/03/130314124616.htm.

7. Haro-Mora, J.J., E. García-Escobar, N. Porras et al. September 1, 2011. "Children Whose Diet Contained Olive Oil Had a Lower Likelihood of Increasing Their Body Mass Index Z-Score Over 1 Year." *European Journal of Endocrinology* 165(3): 435–439.

8. Razquin, C., J.A. Martinez, M.A. Martinez-Gonzalez et al. December 2009. "A 3 Year Follow-Up of a Mediterranean Diet Rich in Virgin Olive Oil Is Associated with High Plasma Antioxidant Capacity and Reduced Body Weight Gain." *European Journal of Clinical Nutrition* 63(12): 1387–1393.

9. Schwartz, Gary J., Jin Fu, Giuseppe Astarita et al. October 8, 2008. "The Lipid Messenger OEA Links Dietary Fat Intake to Satiety." *Cell Metabolism* 8(4): 281–288.

REFERENCES AND RESOURCES

Magazines, Journals, and Newspapers

Benítez-Arciniega, A.D., D. Gómez-Ulloa, A. Vila et al. 2012. "Olive Oil Consumption, BMI, and Risk of Obesity in Spanish Adults." *Obesity Facts* 5(1): 52–59.

Bes-Rastrollo, A. Sánchez-Villegas, C. de la Fuente et al. March 2006. "Olive Oil Consumption and Weight Change: The SUN Prospective Cohort Study." *Lipids* 41(3): 249–256.

Esposito, Katherine, Christina-Maria Kastorini, Demosthenes B. Panagiotakos, and Dario Giugliano. February 2011. "Mediterranean Diet and Weight Loss: Meta-Analysis of Randomized Controlled Trials." *Metabolic Syndrome and Related Disorders* 9(1): 1–12.

Flynn, Mary M. and Steven E. Reinert. June 2010. "Comparing an Olive Oil-Enriched Diet to a Standard Lower-Fat Diet for Weight Loss in Breast Cancer Survivors: A Pilot Study." *Journal of Women's Health* 19(6): 1155–1161.

Haro-Mora, J.J., E. García-Escobar, N. Porras et al. September 1, 2011. "Children Whose Diet Contained Olive Oil Had a Lower Likelihood of Increasing Their Body Mass Index Z-Score Over 1 Year." *European Journal of Endocrinology* 165(3): 435–439.

Razquin, C., J.A. Martinez, M.A. Martinez-Gonzalez et al. December 2009. "A 3 Year Follow-Up of a Mediterranean Diet Rich in Virgin Olive Oil Is Associated with High Plasma Antioxidant Capacity and Reduced Body Weight Gain." *European Journal of Clinical Nutrition* 63(12): 1387–1393.

Schwartz, Gary J., Jin Fu, Giuseppe Astarita et al. October 8, 2008. "The Lipid Messenger OEA Links Dietary Fat Intake to Satiety." *Cell Metabolism* 8(4): 281–288.

Soriguer, F., M.C. Almaraz, M.S. Ruiz-de-Adana et al. November 2009. "Incidence of Obesity Is Lower in Persons Who Consume Olive Oil." *European Journal of Clinical Nutrition* 63(11):1371–1374.

Websites

Olive Oil Source. http://www.oliveoilsource.com.
Science Daily. http://www.sciencedaily.com.

Peanut Oil

While the word "peanut" clearly contains the word "nut," peanuts are not really nuts. They are, in fact, legumes or edible seeds enclosed in pods. They come from the single plant family known as Leguminosae, which includes beans and peas. And, unlike nuts that grow on trees, peanuts grow underground. As a result, peanut oil is not really a nut oil. Nevertheless, in the United States and throughout the entire world, particularly in China, South Asia, and Southeast Asia, peanut oil is used as if it were a nut oil. Furthermore, it is well-known to have a high smoke point, which means that it may be used at very high temperatures.[1] In the United States, peanut oil is often used during Thanksgiving and other holidays to deep fry turkeys.

It is generally believed that the peanut plant originated in South America. Then, European explorers transported the plant to other parts of the world. During the early 1900s, George Washington Carver, who was head of the agriculture department at Tuskegee Institute in Alabama (United States), advised farmers to rotate their cotton crop with other crops, especially peanuts. He noted that unlike cotton, which depletes nitrogen from the soil, peanuts add nitrogen.[2]

According to The Peanut Institute, peanut oil "is one of the healthiest oils." It contains no cholesterol, no trans fats, and very little saturated fat. It is high in

Peanut farmer Tim Burch checks peanuts in
a Baker County, Georgia, field. (AP Photo/
Ric Feld)

unsaturated fats, especially monounsaturated fat. It also contains antioxidants,
such as vitamin E and phytosterols. So, peanut oil should support cardiovascular health. There are three main types—refined, gourmet, and 100% peanut oil.
Refined peanut oil is refined, bleached, and deodorized; during this process, the
peanut oil is rendered nonallergenic. Gourmet peanut oils are not refined; sometimes they are roasted. And, 100% peanut oil contains only peanut oil.[3] It is readily available in supermarkets and specialty stores. Many varieties are sold online,
and they all are reasonably priced.

CARDIOVASCULAR HEALTH

In a study published in 2010 in the *Journal of Food Science*, researchers from
North Carolina (United States) examined the effects of peanut oil, peanuts, and
fat-free peanut flour (FFPF) on cardiovascular disease risk factors and the development of atherosclerosis in male Syrian golden hamsters. The researchers
divided a total of 76 male hamsters into four groups. For 24 weeks, they fed the
hamsters in the control group a high-fat, high-cholesterol diet; the hamsters in
the other groups were fed a similar high-fat, high-cholesterol diet but one group

also included peanut oil, a second group included peanuts, and a third group included FFPF.

The researchers found that the hamsters in all three test groups had significantly lower total plasma cholesterol and low-density lipoprotein (LDL or "bad" cholesterol) than the control hamsters. There were no significant changes in the high-density lipoprotein (HDL or "good" cholesterol). The researchers concluded that peanut oil, peanuts, and FFPF "retarded the development of atherosclerosis in animals consuming an atherosclerosis inducing diet."[4]

In a study published in 2008 in *The Journal of Applied Research,* researchers from Brazil, Ghana, and Indiana (USA) evaluated the effect of the intake of peanut oil on the blood lipid levels of healthy adults between the ages of 18 and 50 years. A total of 64 men and 65 women were assigned to one of four experimental groups. For eight weeks, the members of three of the groups received a daily milk shake containing 30% of their basal energy requirement and either peanut, safflower, or olive oil. The members of the control group did not receive a shake. A number of different tests, such as serum cholesterol levels and blood pressure readings, were conducted at baseline and again at weeks 4 and 8. While safflower oil was found to have stronger LDL-lowering properties than peanut and olive oils, olive oil, and to a lesser extent peanut oil, reduced blood pressure levels. The researchers noted that "the weak and intermediate effects of peanut oil on cholesterol and blood pressure relative to the other oils suggest a contribution of its MUFA [high monounsaturated] composition as well as other constituents in whole peanuts."[5]

IMPROVES THE QUALITY OF THE DIET

In a study published in 2004 in the *Journal of the American College of Nutrition,* researchers from the Pennsylvania State University compared the quality of the diets of men, women, and children who ate peanuts and peanut products to those who did not include peanuts and peanut products in their diets.

Their results were noteworthy. The researchers learned that people who consume peanuts and peanut products have a higher intake of protein, total fat, polyunsaturated fat, monounsaturated fat, fiber, vitamin A, vitamin E, folate, calcium, magnesium, zinc, and iron. Moreover, they have a lower consumption of dietary cholesterol, and they had higher "Health Eating Index" (HEI) ratings. (The HEI was "a measure of overall nutrient profile of the diets.") While the energy intake was high in the peanut users, their body mass index levels were lower. The researchers concluded that "peanuts and peanut products enhance the nutrient profile of the diet."[6]

USEFUL FOR CONTROLLING IODINE DEFICIENCY

In a study published in 2006 in *The American Journal of Clinical Nutrition,* researchers from several countries compared the ability of iodized peanut oil and iodized poppy oil to address iodine deficiency in children, an ongoing challenge in many developing countries. The cohort consisted of 355 schoolchildren between the ages of 8 and 10 who attended one of four different primary schools in Central Java Province, Indonesia. Though goiter is prevalent in this area, only apparently healthy children were included in the study. The children received single oral dose of iodized peanut oil, iodized poppy seed oil, or peanut oil (control). In order to determine the concentrations of iodine, several different measurements of urinary iodine were administered. The researchers found that the participants in the treatment groups had higher concentrations of urinary iodine than those who took the placebo. However, the schoolchildren who took the iodized peanut oil had much higher concentrations of urinary iodine, and these higher concentrations continued for longer periods of time. The researchers concluded that "iodized peanut oil is more efficacious in controlling iodine deficiency than is iodized poppy seed oil containing the same amount of iodine. . . . Iodine retention after a single oral dose of iodized peanut oil was 3 times that after a single oral dose of iodized poppy seed oil, which resulted in double the period of protection in iodine from iodine-deficient schoolchildren."[7]

IS PEANUT OIL BENEFICIAL?

For the vast majority of people, peanut oil does appear to have benefits. Still, an article published in 2012 in the *Journal of the American Academy of Dermatology* noted that "allergic reactions to peanuts in children have become a significant medical and legal concern worldwide, with rising incidence of this potentially fatal condition." So, even tiny amounts of peanuts or nonrefined peanut oil may cause reactions.[8] People with a known allergy to peanuts must be constantly vigilant.

That is why it is particularly disturbing to read the results of a study published in 2012 in the *Annals of Allergy, Asthma & Immunology* on the ability of adults and children to identify peanuts and tree nuts. Researchers in Columbus, Ohio (United States) created a display of 19 peanuts and tree nuts. Well over 1,000 adults and children six years of age or older were asked to identify them. The mean number of peanuts and tree nuts identified by all the subjects was 8.4 or 44.2%. Unfortunately, the mean number of peanuts and tree nuts identified by children was even lower—4.6 or 24.2%. The researchers noted that "both children and adults are unreliable at visually identifying most nuts."[9] Could adults and children differentiate between peanut oil and other similarly appearing oils? It is not clear.

NOTES

1. The Peanut Institute Website. www.peanut-institute.org.
2. Ibid.
3. Ibid.
4. Stephens, Amanda M., Lisa L. Dean, Jack P. Davis et al. May 2010. "Peanuts, Peanut Oil, and Fat Free Peanut Flour Reduced Cardiovascular Disease Risk Factors and the

Development of Atherosclerosis in Syrian Golden Hamsters." *Journal of Food Science* 75(4): H116–H122.

5. Sales, Regiane L., Sandra B. Coelho, Neuza M. B. Costa et al. September 2008. "The Effects of Peanut Oil on Lipid Profile of Normalipidemic Adults: A Three-Country Collaborative Study." *The Journal of Applied Research* 8(3): 216–225.

6. Griel, Amy E., Brenda Eissenstat, Vijaya Juturu et al. December 2004. "Improved Diet Quality with Peanut Consumption." *Journal of the American College of Nutrition* 23(6): 660–668.

7. Untoro, Juliawati, Werner Schultink, Clive E. West et al. November 2006. "Efficacy of Oral Iodized Peanut Oil Is Greater than That of Iodized Poppy Seed Oil among Indonesian Schoolchildren." *The American Journal of Clinical Nutrition* 84(5): 1208–1214.

8. Husain, Zain and Robert A. Schwartz. January 2012. "Peanut Allergy: An Increasingly Common-Life-Threatening Disorder." *Journal of the American Academy of Dermatology* 66(1): 136–143.

9. Hostetler, Todd L., Sarah G. Hostetler, Gary Phillips, and Bryan L. Martin. January 2012. "The Ability of Adults and Children to Visually Identify Peanuts and Tree Nuts." *Annals of Allergy, Asthma & Immunology* 108(1): 25–29.

REFERENCES AND RESOURCES

Magazines, Journals, and Newspapers

Griel, Amy E., Brenda Eissenstat, Vijaya Juturu et al. December 2004. "Improved Diet Quality with Peanut Consumption." *Journal of the American College of Nutrition* 23(6): 660–668.

Hostetler, Todd L., Sarah G. Hostetler, Gary Phillips, and Bryan L. Martin. January 2012. "The Ability of Adults and Children to Visually Identify Peanuts and Tree Nuts." *Annals of Allergy, Asthma & Immunology* 108(1): 25–29.

Husain, Zain and Robert A. Schwartz. January 2012. "Peanut Allergy: An Increasingly Common Life-Threatening Disorder." *Journal of the American Academy of Dermatology* 66(1): 136–143.

Sales, Regiane L., Sandra B. Coelho, Neuza M. B. Costa et al. September 2008. "The Effects of Peanut Oil on Lipid Profile of Normolipidemic Adults: A Three-Country Collaborative Study." *The Journal of Applied Research* 8(3): 216–225.

Stephens, Amanda M., Lisa L. Dean, Jack P. Davis et al. May 2010. "Peanuts, Peanut Oil, and Fat Free Peanut Flour Reduced Cardiovascular Disease Risk Factors and the Development of Atherosclerosis in Syrian Golden Hamsters." *Journal of Food Science* 75(4): H116–H122.

Untoro, Juliawati, Werner Schultink, Clive E. West et al. November 2006. "Efficacy of Oral Iodized Peanut Oil Is Greater than That of Iodized Poppy Seed Oil among Indonesian School Children." *The American Journal of Clinical Nutrition* 84(5): 1208–1214.

Website

The Peanut Institute. www.peanut-institute.org.

Peppermint Oil

Grown in North America and Europe, peppermint is a member of the mint family. Most likely, the medicinal use of peppermint may be traced to ancient Greece, where it was used as a digestive aide and to manage gallbladder disease. It was also considered an effective treatment for upper respiratory problems.

Extracted from the stem, leaves, and flowers of the plant, peppermint oil is believed to be an effective treatment for irritable bowel syndrome and nonulcer dyspepsia (indigestion). When applied topically, it is thought to ease headache symptoms.[1]

IRRITABLE BOWEL SYNDROME

In a randomized, double-blind, placebo-controlled study published in 2007 in *Digestive and Liver Disease*, researchers from Italy wanted to learn if enteric-coated peppermint capsules would help people dealing with irritable bowel syndrome, a disorder characterized by recurrent episodes of abdominal distension and bloating, abdominal pain, and altered bowel habits with constipation, diarrhea, and an urgency to defecate. Their initial cohort consisted of 57 people. For four weeks, 28 people took two enteric-coated peppermint oil capsules twice daily and 29 took the same number of placebos. The participants were evaluated at the

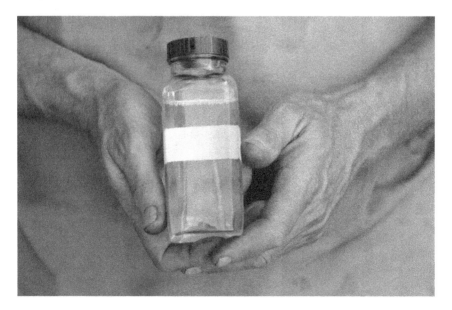

Mint grower Larry Wappell holds a sample bottle of peppermint oil on his farm in San Pierre, Indiana. (AP Photo/Michael Conroy)

end of the study and, again, four weeks after the study concluded. The research-ers found that the participants in the peppermint oil group had a greater than 50% reduction in irritable bowel syndrome symptoms; this contrasted to the 38% reduction in the placebo group. The researchers commented that "treat-ment with enteric-coated capsules of peppermint oil given twice a day for 4 weeks is more effective than placebo in reducing abdominal symptoms related to IBS." In addition, "the beneficial effect of peppermint oil also lasts for 1 month after therapy in more than 50% of treated patients."[2]

A few years later, in 2010, a study published in *Digestive Diseases and Sci-ences* also examined the use of peppermint oil for irritable bowel syndrome. In this randomized, double-blind, placebo-controlled trial, researchers from Iran placed 90 participants in one of two groups. The participants in one group took a daily capsule of peppermint oil (brand name Colpermin); the partici-pants in the other group took a daily placebo. At the outset of the study, the most common complaint was abdominal pain. That was followed by distension and flatulence. The participants were evaluated after one week, four weeks, and eight weeks, when the study concluded. Sixty people completed the study. The researchers found that the participants "on Colpermin, as compared to controls, showed statistically significant improvement." However, while they observed improvement in abdominal pain and discomfort, there was no re-lief for the symptoms of diarrhea, constipation, or bloating. The researchers concluded "that treatment with Colpermin may be effective in alleviating abdominal pain and discomfort and improving some aspects of quality of life in patients with IBS."[3]

In an analysis published in 2010 in the *European Journal of Gastroenterology & Hepatology*, researchers from Germany examined the effectiveness of 121 trials published over 35 years on treatments offered to people with irritable bowel syn-drome. They found considerable variation in the ranking of the different treat-ment options. Peppermint oil had the highest rating. "Best efficiency is currently available for herbal intervention (peppermint oil) followed by psychological and psychopharmacological therapies and by quite old substances such as spasmolyt-ics and probiotics."[4]

In another analysis, published in 2008 in *BMJ*, researchers from Canada, Ire-land, and several locations in the United States reviewed studies that examine the usefulness of fiber-rich foods, antispasmodics, and peppermint oil for the symptoms of irritable bowel syndrome. They found that all three treatments were useful. "This systematic review and meta-analysis has shown that fibre, antispas-modics, and peppermint oil are all more effective than placebo in the treatment of irritable bowel syndrome."[5]

MAY BE USEFUL DURING ENDOSCOPIC PROCEDURES

In a nonrandomized, prospective study published in 2012 in *Digestive Diseases and Sciences*, researchers from Japan investigated the use of peppermint oil as an antispasmodic drug during endoscopic procedures designed to see the upper portion of the gastrointestinal tract, a procedure that is scientifically known as esophagogastroduodenoscopy (EGD). Why is there a need for another form of antispasmodic medication? The current medications have side effects. For example, hyoscine butyl bromide (HB) may cause palpitations, dry mouth, and urinary retention. And, glucagon (GL) may cause hyperglycemia and reactive hypoglycemia.

The cohort consisted of 8,269 people who all had EGD procedures. Of these, peppermint oil was used in 1,893, HB was used in 6,063, and GL was used in 157. One hundred and fifty-six patients had no antispasmodic medication. The researchers learned that peppermint oil was an effective antispasmodic agent, especially for those who were 70 years of age and older. "Peppermint oil was useful as an antispasmodic drug during EGD, especially for elderly patients."[6]

In a randomized, placebo-controlled study published in 2011 in *Clinical Pharmacology & Therapeutics*, researchers from Tokyo, Japan, wanted to learn if peppermint oil could reduce gastric spasms that may interfere with an endoscopy of the upper gastrointestinal tract. The initial cohort consisted of 24 healthy Japanese male volunteers between the ages of 35 and 64. They were placed in groups that were treated with different doses of L-menthol, a component of peppermint oil, or a placebo. (The L-menthol was sprayed directly on the gastric mucosa.) The researchers noted that the L-menthol was rapidly absorbed. First levels were detectable at five minutes; peak concentrations were reached within one hour. And, the L-menthol, in doses as high as 320mg, "was well tolerated." When compared to baseline, L-menthol tended to disease the occurrence of gastric spasms, but the results were not statistically significant.[7]

A description of Phase 2 of this study was published in 2012 in *Digestive Endoscopy*. During this second phase, researchers used a video-recording endoscope to assess the degree that L-menthol was able to relieve gastric spasms. In addition, they evaluated the safety of L-menthol, and they quantified the gastric responses to different doses of L-menthol. The cohort consisted of 131 subjects; of these, 116 received various doses of L-menthol. The median age was 60.5 years. The researchers noted that "increasing doses of L-menthol sprayed on the gastric mucosa were associated with significant increases in the peristalsis-suppressing effect during upper GI endoscopy." And, they concluded that "L-menthol reduces gastric peristalsis in a dose-dependent manner, and the dose-response reaches a plateau at 0.8%."[8]

MAY BE USEFUL FOR MIGRAINES

In a study published in 2010 in the *International Journal of Clinical Practice*, researchers from Iran wanted to determine if the application of a 10% solution of menthol to the forehead would provide relief from the painful symptoms of migraine headache. Though the initial cohort consisted of 35 people, only 25 people completed the study and were included in the results. When compared to the placebo, the researchers found that menthol in ethanol result in a significant reduction or elimination of migraine pain. The menthol solution "was also more efficacious in [the] alleviation of associated symptoms" such as nausea and vomiting. The researchers concluded that they demonstrated "the efficacy, safety and relative tolerability of menthol solution for the abortive treatment of migraine without aura."[9]

IS PEPPERMINT OIL BENEFICIAL?

It is well known that millions of people deal with the symptoms of irritable bowel syndrome. For them, it may be useful to try peppermint oil for a few weeks. To prevent the release of the oil into the stomach, it is better to purchase enteric-coated peppermint oil capsules. The capsules should remain intact until they reach the intestines. Enteric-coated peppermint oil capsules should never be taken with a medication that reduces stomach acid, such as Pepcid. The medication may cause the peppermint oil to dissolve in the stomach before reaching the intestines. And, the millions of people who suffer from migraines may also want to try dabbing a little peppermint oil on their foreheads. Before beginning a regime of peppermint oil, it is a good idea to discuss your plans with a medical provider.

NOTES

1. Kligler, Benjamin and Sapna Chaudhary. April 1, 2007. "Peppermint Oil." *American Family Physician* 75(7): 1027–1030.

2. Cappello, G., M. Spezzaferro, L. Grossi et al. June 2007. "Peppermint Oil (Mint-oil) in the Treatment of Irritable Bowel Syndrome: A Prospective Double-Blind Placebo-Controlled Randomized Trial." *Digestive and Liver Disease* 39(6): 530–536.

3. Merat, S., S. Khalili, P. Mostajabi et al. May 2010. "The Effect of Enteric-Coated Delayed-Release Peppermint Oil on Irritable Bowel Syndrome." *Digestive Diseases and Sciences* 55(5): 1385–1390.

4. Enck, P., F. Junne, S. Klosterhalfen et al. December 2010. "Therapy Options in Irritable Bowel Syndrome." *European Journal of Gastroenterology & Hepatology* 22(12): 1402–1411.

5. Ford, A.C., N.J. Talley, B.M. Spiegel at al. November 13, 2008. "Effect of Fibre, Antispasmodics, and Peppermint Oil in the Treatment of Irritable Bowel Syndrome: Systematic Review and Meta-Analysis." *BMJ* 337: a2313.

6. Imagawa, A., H. Hata, N. Nakatsu et al. September 2012. "Peppermint Soil Solution Is Useful as an Antispasmodic Drug for Esophago Gastro Duodeno Scopy, Especially for Elderly Patients." *Digestive Diseases and Sciences* 57(9): 2379–2384.

7. Hiki, N., M. Kaminishi, T. Hasunuma et al. 2011. "A Phase 1 Study Evaluating Tolerability, Pharmacokinetics, and Preliminary Efficacy of L-Menthol in Upper Gastrointestinal Endoscopy." *Clinical Pharmacology &Therapeutics &*90(2): 221–228.

8. Hiki, N., M. Kaminishi, K. Yasuda et al. March 2012. "Multicenter Phase II Randomized Study Evaluating Dose-Response of Antiperistaltic Effect of L-Menthol Sprayed onto the Gastric Mucosa for Upper Gastrointestinal Endoscopy." *Digestive Endoscopy* 24(2): 79–86.

9. Borhani, Haghighi, A., S. Motazedian, R. Resaii et al. March 2010. "Cutaneous Application of Menthol 10% Solution as an Abortive Treatment of Migraine, without Aura: A Randomised, Double-Blind, Placebo-Controlled, Cross-Over Study." *International Journal of Clinical Practice* 64(4): 451–456.

REFERENCES AND RESOURCES

Magazines, Journals, and Newspapers

Borhani Haghighi, A., S. Motazedian, R. Resaii et al. March 2010. "Cutaneous Application of Menthol 10% Solution as an Abortive Treatment of Migraine without Aura: A Randomised, Double-Blind, Placebo-Controlled Cross-Over Study." *International Journal of Clinical Practice* 64(4): 451–456.

Cappello, G., M. Spezzaferro, L. Grossi et al. June 2007. "Peppermint Oil (Mintoil) in the Treatment of Irritable Bowel Syndrome: A Prospective Double-Blind Placebo-Controlled Randomized Trial." *Digestive and Liver Disease* 39(6): 530–536.

Enck, P., F. Junne, S. Klosterhalfen et al. December 2010. "Therapy Options in Irritable Bowel Syndrome." *European Journal of Gastroenterology & Hepatology* 22(12): 1402–1411.

Ford, A. C., N. J. Talley, B. M. Spiegel et al. November 13, 2008. "Effect of Fibre, Antispasmodics, and Peppermint Oil in the Treatment of Irritable Bowel Syndrome: Systematic Review and Meta-Analysis." *BMJ* 337: a2313.

Hiki, N., M. Kaminishi, K. Yasuda et al. March 2012. "Multicenter Phase II Randomized Study Evaluating Dose-Response of Antiperistaltic Effect of L-Menthol Sprayed onto the Gastric Mucosa for Upper Gastrointestinal Endoscopy." *Digestive Endoscopy* 24(2): 79–86.

Hiki, N., M. Kaminishi, T. Hasunuma et al. 2011. "A Phase 1 Study Evaluating Tolerability, Pharmacokinetics, and Preliminary Efficacy of L-Menthol in Upper Gastrointestinal Endoscopy." *Clinical Pharmacology & Therapeutics* 90(2): 221–228.

Imagawa, A., H. Hata, M. Nakatsu et al. September 2012. "Peppermint Oil Solution Is Useful as an Antispasmodic Drug for Esophago Gastro Duodeno Scopy, Especially for Elderly Patients." *Digestive Diseases and Sciences* 57(9): 2379–2384.

Kligler, Benjamin and Sapna Chaudhary. April 1, 2007. "Peppermint Oil." *American Family Physician* 75(7): 1027–1030.

Merat, S., S. Khalili, P. Mostajabi et al. May 2010. "The Effect of Enteric-Coated Delayed-Release Peppermint Oil on Irritable Bowel Syndrome." *Digestive Diseases and Sciences* 55(5): 1385–1390.

Website

University of Maryland Medical Center. www.umm.edu.

Pumpkin Seed Oil

Many people think of pumpkins, pumpkin seeds, and pumpkin seed oil only in association with Halloween and other holidays and special events in the fall. But, these pumpkin products are actually much more than seasonal products. They are all incredibly healthful foods that may be enjoyed throughout the entire year. Specifically, pumpkin seed oil contains zinc, selenium, phytoestrogens, and vitamins A and E. It has the essential fatty acids omega-3 and omega-6. And, it is filled with antioxidants that destroy free radicals.[1]

Expect pumpkin seed oil to be a little difficult to find in traditional markets. It may be more readily available in specialty stores and online. The prices appear to vary significantly.

MAY SUPPORT PROSTATE HEALTH

In a randomized, double-blind, placebo-controlled trial published in 2009 in *Nutrition Research and Practice*, researchers from Korea investigated the use of pumpkin seed oil to treat men with benign prostatic hyperplasia—enlargement of the prostate gland—a common condition in older men. Sixty-two men with benign prostatic hyperplasia were divided into four groups. During the course of 12 months, the subjects took capsules of sweet potato starch (control), pumpkin seed oil (320mg/day), saw palmetto oil (320mg/day), or the combination of pumpkin seed oil (320mg/day) and saw palmetto oil (320mg/day). Forty-seven men completed the trial; their average age was 53.3 years. When compared to the men in the control group, the men in the three treatment groups experienced improvement in their symptoms. The researchers concluded that "based on the results of this study, it could be suggested that pumpkin seed oil and saw palmetto oil are clinically safe and may be effective complementary and alternative medicine for BPH [benign Prostatic hyperplasia]."[2]

In a study published in 2006 in *Urologia Internationalis*, researchers from Taiwan examined the ability of pumpkin seed oil and Phytosterol-F, a plant-based sterol, to control testosterone/prazosin-induced prostate growth in rats. (These rats are hormonally and chemically induced to have excessive prostate growth.) The researchers began by dividing 40 adult Wistar rats into five groups. While one group of rats was set aside to serve as the control, the other rats were fed various combinations of testosterone/prazosin, pumpkin seed oil, and Phytosterol-F. The results are important. When compared to the rats that were only treated with testosterone/prazosin, the rats treated with testosterone/prazosin and pumpkin seed oil or testosterone/prazosin and pumpkin seed oil and Phytosterol-F had significantly reduced prostate gland growth.[3]

In a study published in 2006 in the *Journal of Medicinal Food*, researchers from Jamaica used the subcutaneous administration of testosterone to enlarge the prostate glands of rats. Then, for the next 20 days, the rats were fed one of two doses of either pumpkin seed oil or corn oil. On the 21st day of the trial, the rats were sacrificed, and their prostates were removed. According to the researchers, "testosterone significantly increased prostate size ratio . . . and this induced increase was inhibited in rats fed pumpkin seed oil." They concluded that pumpkin seed oil "can inhibit testosterone-induced hyperplasia of the prostate and therefore may be beneficial in the management of benign prostate hyperplasia."[4]

Interestingly, in a Letter to the Editor of the *West Indian Medical Journal*, which was published in 2008, the authors indicated that patients visiting a urology clinic from October 2005 to March 2006 were asked what prescribed and herbal medications they were taking for their enlarged prostate glands. Of the 65 men diagnosed with enlarged prostate glands, 18 noted that they were taking herbal preparations. The most popular choices were products containing pumpkin seeds, aloe vera, and saw palmetto. "The data collected suggest that there may be a high prevalence of the use of herbal products among patients being managed for BPH and most users appear to be convinced of their benefits."[5]

MAY SUPPORT CARDIOVASCULAR HEALTH

In a study published in 2012 in the *Journal of Medicinal Food*, researchers from Cairo, Egypt, began by inducing hypertension (high blood pressure) in rats. For six weeks, some rats received 40 or 100mg/kg of pumpkin seed oil; others received amlodipine (0.9mg/kg), a calcium channel blocker medication used to treat hypertension. Still, other rats served as the controls. Interestingly, the researchers found that both pumpkin seed oil and amlodipine significantly decreased the elevated blood pressure levels of the rats. The researcher concluded that their findings have "shown that pumpkin seed oil exhibits antihypertensive and cardioprotective effects."[6]

In a study published in 2008 in *Phytotherapy Research*, researchers from Jamaica investigated the effect of pumpkin seed oil supplementation on several markers of cardiovascular health, such as total cholesterol and low density lipoprotein (LDL or "bad" cholesterol) in nonovariectomized and ovariectomized Sprague-Dawley rats. The control rats were fed corn oil. For 12 weeks, the rats were fed the oils five days per week—Monday to Friday. The results were notable. Pumpkin seed oil supplementation "produced positive influences on the plasma lipid profile of both non-ovariectomized and ovariectomized female rats." Specially, there were improvements in cholesterol levels and blood pressure readings. As a result, the researchers hypothesized that menopausal women who supplement their diets with pumpkin seed oil may be able to reduce their risk of the "cardiovascular complications" that are associated with dramatically lower levels of estrogen.[7]

MAY HELP PEOPLE DEAL WITH SIDE EFFECTS
OF METHOTREXATE TREATMENT

Methotrexate is a medication that is widely used to treat various types of cancer as well as inflammatory illnesses such as arthritis. However, it is known to have a number of problematic side effects. One of the most frequent is entero-colitis, inflammation of the colon or small intestine. In a study published in 2011 in the *Indian Journal of Biochemistry & Biophysics*, a researcher from Egypt wanted to learn if pumpkin seed oil and ellagic acid (a phytochemical found in plant foods such as raspberries and strawberries) could protect against this side effect.

So, the researchers randomly divided 40 rats into five groups of 8 rats. The first group of rats, which were the controls, was injected with saline. The second group of rats was injected with methotrexate. The third group of rats was injected with methotrexate, but also treated with pumpkin seed oil dissolved in dimethyl sulphoxide. (Dimethyl sulphoxide or DMSO is a widely used solvent in biological studies.) The rats in the fourth group were injected with methotrexate and also received ellagic acid dissolved in DMSO. The fifth group of rats was in-jected with methotrexate and orally treated with DMSO. The researcher found that both pumpkin seed oil and ellagic acid reduced the amount of damage to the small intestine caused by the methotrexate. When DMSO is used alone, the damage caused by methotrexate is only "slightly ameliorated." The researcher concluded that "the present study demonstrated the evidence of intestinal dam-age by MTX [methotrexate] and ability of PSO [pumpkin seed oil] and EA [ellagic acid] in preventing the damage."[8]

MAY BE USEFUL FOR POSTMENOPAUSAL WOMEN

In a randomized, double-blinded, placebo-controlled study published in 2011 in *Climacteric: The Journal of the International Menopausal Society*, researchers from Jamaica wanted to determine if pumpkin seed oil would assist women dealing with the symptoms of menopause. The cohort consisted of 35 women who had undergone menopause. Over a 12-week period, 21 women took 2g/day of pump-kin seed oil and 14 women took 2g/day of wheat germ oil (control).

The researchers found that the women taking pumpkin seed oil had significant increases in high density lipoprotein (HDL or "good" cholesterol) concentrations and decreases in their diastolic blood pressure (bottom number). They also ex-perienced significant improvements in their menopausal symptoms—fewer hot flashes and headaches and less joint pain. On the other hand, the researchers commented that the women taking wheat germ oil "reported being more de-pressed and having more unloved feeling." They concluded that "pumpkin seed oil had some benefits for postmenopausal women and provided strong evidence to support further studies."[9]

IS PUMPKIN SEED OIL BENEFICIAL?

While the research on pumpkin seed oil appears to be somewhat limited, it is apparent that it may well be useful for several medical concerns.

NOTES

1. Pumpkin Seed Oil Website. www.pumpkinseedoil.com.

2. Hong, Heeok, Chun-Soo Kim, and Sungho Maeng. Winter 2009. "Effects of Pumpkin Seed Oil and Saw Palmetto Oil in Korean Men with Symptomatic Benign Prostatic Hyperplasia." *Nutrition Research and Practice* 3(4): 323–327.

3. Tsai, Y. S., Y. C. Tong, J. T. Cheng et al. October 2006. "Pumpkin Seed Oil and Phytosterol-F Can Block Testosterone/Prazosin-Induced Prostate Growth in Rats." *Urologia Internationalis* 77(3): 269–274.

4. Gossell-Williams, M., A. David, and N. O'Connor. Summer 2006. "Inhibition of Testosterone-Induced Hyperplasia of the Prostate of Sprague-Dawley Rats by Pumpkin Seed Oil." *Journal of Medicinal Food* 9(2): 284–286.

5. Gossell-Williams, M., A. Davis, W. Aiken, and R. Mayhew. January 2008. "Herbal Preparation Use among Patients with Benign Prostatic Hyperplasia Attending a Urology Clinical in Jamaica, West Indies." *West Indian Medical Journal* 57(1): 75, 76.

6. El-Mosallamy, A. E., A. A. Sleem, O. M. Abdel-Salam et al. February 2012. "Antihypertensive and Cardioprotective Effects of Pumpkin Seed Oil." *Journal of Medicinal Food* 15(2): 180–189.

7. Gossell-Williams, M., K. Lyttle, T. Clarke et al. July 2008. "Supplementation with Pumpkin Seed Oil Improves Plasma Lipid Profile and Cardiovascular Outcomes of Female Non-Ovariectomized and Ovariectomized Sprague-Dawley Rats." *Phytotherapy Research* 22(7): 873–877.

8. El-Boghdady, Noha A. 2011. "Protective Effect of Ellagic Acid and Pumpkin Seed Oil against Methotrexate-Induced Small Intestine Damage in Rats." *Indian Journal of Biochemistry & Biophysics* 48(6): 380–387.

9. Gossell-Williams, M., C. Hyde, T. Hunter et al. October 2011. "Improvement in HDL Cholesterol in Postmenopausal Women Supplemented with Pumpkin Seed Oil: Pilot Study." *Climacteric: The Journal of the International Menopause Society* 14(5): 558–564.

REFERENCES AND RESOURCES

Magazines, Journals, and Newspapers

El-Boghdady, Noha A. December 2011. "Protective Effect of Ellagic Acid and Pumpkin Seed Oil against Methotrexate-Induced Small Intestine Damage in Rats." *Indian Journal of Biochemistry & Biophysics* 48(6): 380–387.

El-Mosallamy, A. E., A. A. Sleem, O. M. Abdel-Salam et al. February 2012. "Antihypertensive and Cardioprotective Effects of Pumpkin Seed Oil." *Journal of Medicinal Food* 15(2): 180–189.

Gossell-Williams, M., A. Davis, and N. O'Connor. Summer 2006. "Inhibition of Testosterone-Induced Hyperplasia of the Prostate of Sprague-Dawley Rats by Pumpkin Seed Oil." *Journal of Medicinal Food* 9(2): 284–286.

Gossell-Williams, M., A. Davis, W. Aiken, and R. Mayhew. January 2008. "Herbal Preparation Use among Patients with Benign Prostatic Hyperplasia Attending a Urology Clinic in Jamaica, West Indies." *West Indian Medical Journal* 57(1): 75, 76.

Gossell-Williams, M., C. Hyde, T. Hunter et al. October 2011. "Improvement in HDL Cholesterol in Postmenopausal Women Supplement with Pumpkin Seed Oil: Pilot Study." *Climacteric: The Journal of the International Menopause Society* 14(5): 558–564.

Gossell-Williams, M., K. Lyttle, T. Clarke et al. July 2008. "Supplementation with Pumpkin Seed Oil Improves Plasma Lipid Profile and Cardiovascular Outcomes of Female Non-Ovariectomized and Ovariectomized Sprague-Dawley Rats." *Phytotherapy Research* 22(7): 873–877.

Hong, Heeok, Chun-Soo Kim, and Sungho Maeng. Winter 2009. "Effects of Pumpkin See Oil and Saw Palmetto Oil in Korean Men with Symptomatic Benign Prostatic Hyperplasia." *Nutrition Research and Practice* 3(4): 323–327.

Tsai, Y.S., Y.C. Tong, J.T. Cheng et al. October 2006. "Pumpkin Seed Oil and Phytosterol-F Can Block Testosterone/Prazosin-Induced Prostate Growth in Rats." *Urologia Internationalis* 77(3): 269–274.

Website

Pumpkin Seed Oil. www.pumpkinseedoil.com.

Red Palm Oil

Believed to originate in West Africa, red palm oil, which is produced from the fruit of the *Elaeis guineensis* (African oil palm) tree, has truly ancient roots. The Pharaohs of ancient Egypt considered it a sacred food. And, when they died, it was entombed with them so that they could take the oil into the afterlife.

The fruits from which the Africans produce their palm oil. (iStockPhoto)

More recently, red palm oil has been used as an all-purpose remedy for a wide variety of medical problems. This is probably because it is filled with a wide variety of vitamins, antioxidants, and other phytonutrients. In fact, it is likely the most nutrient-packed dietary oil. For example, the red color of red palm oil is derived from carotenes, such as beta-carotene. The body converts them into vitamin A. A deficiency of vitamin A may cause blindness, impaired learning, weakened bones, and lowered immunity. While vitamin A deficiency is rare in the developed world, it is not uncommon in parts of the developing world. Red palm oil also has very high amounts of the antioxidant vitamin E, which is believed to protect against aging, cardiovascular disease, cancer, arthritis, and Alzheimer's disease. It is frequently viewed as an essential food for expectant and nursing moms. Many believe that it supports the health of both the mother and child.[1]

During the past few years, increasing numbers of people have been learning about the benefits of red palm oil. But, it is still hard to purchase in supermarkets. It is easier to locate online, but it is often expensive.

IMPROVES CARDIOVASCULAR HEALTH

In a study published in 2011 in *Lipids in Health and Disease,* researchers from Hungary and South Africa examined the ability of red palm oil supplementation to effect infarct size in rats. (Infarct size is the amount of damage following a heart attack.) For nine weeks, rats were fed either a normal rat diet or a rat diet enriched with 2% cholesterol. Beginning with week 4, some of the rats also received red palm oil supplementation. The rats were then subjected to 30 minutes of reduced blood flow to the heart (ischemia), which caused heart damage. Next, the rats received two hours of blood flowing back into the heart (reperfusion). After the animals were sacrificed, the researchers learned that the red palm oil significantly reduced the amount of heart damage. According to the researchers, red palm oil supplementation "reversed the negative effects of cholesterol supplementation in the ischaemia/reperfusion heart model." In addition, red palm oil supplementation "reduced myocardial infarct size in cholesterol supplemented rats."[2]

In a study published in 2008 in the Spanish language *Investigación Clínica,* researchers based in Venezuela evaluated the effect of red palm oil on the lipid profile and levels of vitamins A and E in four groups of rats. One group of rats was fed standard rat food; this group served as the control. A second group of rats was fed the standard food enriched with egg yoke powder. A third group of rats was fed the standard food enriched with egg yoke powder and red palm oil. (Yoke powder was used to elevate lipid levels.) The final group was fed the standard food enriched with red palm oil. After 35 days, the researchers found that, when compared to the control group, the red palm oil significantly lowered total cholesterol levels in the rats in the third and fourth groups. The rats in the third and fourth groups also had significant increases in vitamins A and E. To the researchers, red palm oil appears to lower levels of cholesterol and to be rich in a number of desirable nutrients.[3]

In a study published in 2005 in the *Journal of Nutritional Biochemistry*, researchers from Massachusetts, Pennsylvania (both United States), and Malaysia examined the use of palm oil products on hamsters with elevated levels of cholesterol. Forty-eight hamsters were divided into four groups of 12. For two weeks, all the hamsters were fed a high-cholesterol diet. Then, for 10 weeks, the hamsters in the first group continued to eat a high-cholesterol diet, the hamsters in the second group ate a high-cholesterol diet with red palm oil, the hamsters in the third group at a high-cholesterol diet and a different palm oil product, and the hamsters in the fourth group ate a high-cholesterol diet and still another palm oil product. The researchers found that all three palm oil products, including red palm oil, lowered levels of total cholesterol and raised levels of high density lipoprotein (HDL or "good" cholesterol).[4]

In a study published in 2011 in *Food and Nutrition Sciences*, researchers from Malaysia tested four different vegetables oil, including red palm oil, on the lipid profiles of rats. They began by dividing 66 rats into 11 groups of 6 rats. The rats in one group served as the controls. Rats in the other groups received supplementation with the different oils for four and eight weeks. After four weeks, the rats receiving red palm oil supplementation experienced a lowering of their low-density lipoprotein (LDL or "bad" cholesterol). The red palm oil also raised the HDL levels. At eight weeks, the rats fed the red palm oil supplementation had significant decreases in their LDL. The researchers concluded that red palm oil resulted in increases in HDL and decreases in LDL. So, it clearly has properties that support cardiovascular health.[5]

IMPROVES BLOOD CONCENTRATIONS
OF IMPORTANT NUTRIENTS

In a study published in 2003 in *Biomedical and Environmental Sciences*, researchers from Beijing, China, wanted to learn some of the effects of consuming red palm oil in 42 Chinese male soldiers between the ages of 18 and 32. Twenty men from one army company were assigned to consume red palm oil with their diet, and 22 men from another army company were assigned to consume soybean oil. The volunteers consumed these oils as part of their diets for six weeks. At the end of the trial, the researchers observed that the men who consumed red palm oil had increased levels of carotenoids and vitamin E as well as increases in blood concentrations of important nutrients, such as lycopene.[6]

In a study published in 2006 in *Nutrition Journal*, researchers from Canada and Burkina Faso, a relatively poor country in West Africa, assessed the impact of adding red palm oil to the school lunch program. The researchers explained that there is widespread vitamin A deficiency among the children of this sub-Saharan country. "In a small community-based study conducted in 1999 in the north-central part of Burkina Faso, 84.5% of under-five children and 61.8% of their mothers were VA [vitamin A]-deficient according to serum retinol concentrations." So, during one school year, red palm oil was added to individual meals three times a week in certain primary schools. The researchers tested for levels of vitamin in the blood at baseline and 12 months after the trial began. As expected, the researchers found high rates of vitamin A deficiency in the children they tested. Supplementation with red palm oil proved to be an excellent source of vitamin A. "RPO [red palm oil] is a highly bio-effective source of VA." And, apparently, the children truly enjoyed the red palm oil. "It was very popular among exposed pupils in this study."[7]

IS RED PALM OIL BENEFICIAL?

From the available research, red palm oil seems to have a number of different positive properties. But, the research is somewhat limited. And, there is very little available research on the consumption of red palm oil by humans. Hopefully, there will be more research on this interesting oil in the future.

NOTES

1. Oguntibeju, O.O., A.J. Esterhuyse, and E.J. Truter. 2009. "Red Palm Oil: Nutritional, Physiological, and Therapeutic Roles in Improving Human Wellbeing and Quality of Life." *British Journal of Biomedical Science* 66(4): 216–222.

2. Szucs, Gergo, Dirk J. Bester, Krisztina Kupai et al. 2011. "Dietary Red Palm Oil Supplementation Decreases Infarct Size in Cholesterol Fed Rats." *Lipids in Health and Disease* 10: 103+.

3. Salinas, N., M. Márquez, R. Sutil et al. March 2008. "Effect of Partially Refined Palm Oil in Lipid Profile in Rats." *Investigación Clínica* 49(1): 5–16.

4. Wilson, Thomas A., Robert J. Nicolosi, Timothy Kotyla et al. 2005. "Different Palm Oil Preparations Reduce Plasma Cholesterol Concentrations and Aortic Cholesterol Accumulation Compared to Coconut Oil in Hypercholesterolemic Hamsters." *Journal of Nutritional Biochemistry* 16(10): 633–640.

5. Dauqan, Eqbal, Halimah Abdullah Sani, Aminah Abdullah, and Zalifah Mohd Kasim. 2011. "Effect of Different Vegetable Oils (Red Palm Olein, Palm Olein, Corn Oil and Coconut Oil) on Lipid Profile in Rat." *Food and Nutrition Sciences* 2(4): 253–258.

6. Zhang, Jian, Chun-Rong Wang, An-Na Xue, and Ke-You Ge. 2003. "Effects of Red Palm Oil on Serum Lipids and Plasma Carotenoids Level in Chinese Male Adults." *Biomedical and Environmental Sciences* 16(4): 348–354.

7. Zeba, Augustin N., Yves Martin Prével, Issa T. Somé, and Hélène F. Delisle. 2006. "The Positive Impact of Red Palm Oil in School Meals on Vitamin A Status: Study in Burkina Faso." *Nutrition Journal* 5(1): 17.

REFERENCES AND RESOURCES
Magazines, Journals, and Newspapers

Dauqan, Eqbal, Halimah Abdullah Sani, Aminah Abdullah, and Zalifah Mohd Kasim. 2011. "Effect of Different Vegetable Oils (Red Palm Olein, Palm Olein, Corn Oil and Coconut Oil) on Lipid Profile in Rat." *Food and Nutrition Sciences* 2(4): 253–258.

Oguntibeju, O.O., A.J. Esterhuyse, and E.J. Truter. 2009. "Red Palm Oil: Nutritional, Physiological and Therapeutic Roles in Improving Human Wellbeing and Quality of Life." *British Journal of Biomedical Science* 66(4): 216–222.

Salinas, N., M. Márquez, R. Sutil et al. March 2008. "Effect of Partially Refined Palm Oil in Lipid Profile in Rats." *Investigación Clínica* 49(1): 5–16.

Szucs, Gergo, Dirk J. Bester, Krisztina Kupai et al. 2011. "Dietary Red Palm Oil Supplementation Decreases Infarct Size in Cholesterol Fed Rats." *Lipids in Health and Disease* 10: 103.

Wilson, Thomas A., Robert J. Nicolosi, Timthy Kotyla et al. 2005. "Different Palm Oil Preparations Reduce Plasma Cholesterol Concentrations and Aortic Cholesterol Accumulation Compared to Coconut Oil in Hypercholesterolemic Hamsters." *Journal of Nutritional Biochemistry* 16(10): 633–640.

Zeba, Augustin N., Yves Martin Prével, Issa T. Somé, and Hélène F. Delisle. 2006. "The Positive Impact of Red Palm Oil in School Meals on Vitamin A Status: Study in Burkina Faso." *Nutrition Journal* 5(1): 17.

Zhang, Jian, Chun-Rong Wang, An-Na Xue, and Ke-You Ge. 2003. "Effects of Red Palm Oil on Serum Lipids and Plasma Carotenoids Level in Chinese Male Adults." *Biomedical and Environmental Sciences* 16(4): 348–354.

Website

Palm Oil Health. www.palmoilhealth.org.

Rice Bran Oil

Also known as rice bran extract, rice bran oil is produced from the bran of the rice kernel. Known for its mild, nut-like flavor, it contains a number of substances that are thought to be healthful, including gamma-oryzanol, tocopherols, and tocotrienols.[1] It is primarily believed to support cardiovascular health. However, it may be effective for other concerns.

Years ago, rice bran oil was not readily available for sale. Today, it is relatively easy to find, especially in health food stores and online. Suitable for high-temperature cooking, rice bran oil is moderately priced. It is a popular oil in several Asian countries such as Japan and China. And, it is gaining popularity in the United States.

CARDIOVASCULAR HEALTH

In a double-blind, controlled, parallel group study published in 2010 in the *Journal of the Indian Medical Association*, researchers based in India wanted to learn if a combination of rice bran oil and safflower oil would help people lower their elevated levels of cholesterol. The cohort consisted of 73 people who were randomly asked to switch to the rice bran/safflower blended oil or to continue to use the oil that they normally used. Lipid profiles were monitored monthly.

The researchers noted that "at each follow-up, LDL-C [low-density lipoprotein cholesterol or "bad" cholesterol] levels showed a significant reduction from baseline in the study oil group and [the] reduction was more than that observed in the control group." By the end of the three-month study, 82% of the people who consumed the blended oil had LDL levels of less than 150mg/dL compared to 57% in the control group. The researchers commented that "the substitution of usual cooking oil with a blend of rice brand oil and safflower oil (8:2) was found to exert beneficial effects on the LDL-C levels shifting them to low-risk lipid category."[2]

In a 48-week study published in 2009 in *The Journal of International Medical Research*, researchers from Thailand wanted to learn if rice bran oil, soybean oil, palm oil, or a combination of rice bran oil and palm oil would be useful for women with elevated levels of LDLs. The cohort consisted of 16 women between the ages of 44 and 67 with elevated levels of total cholesterol and LDL cholesterol. The study began with an eight-week weight-maintaining diet. That was followed by four 10-week study periods in which the subjects ate the same diet supplemented with one of the three test oils or the combination of rice bran oil and palm oil. The researchers found that the consumption of rice bran oil, soybean oil, and the combination of rice bran oil and palm oil significantly reduced levels of total cholesterol and LDL.[3]

In a randomized, crossover study published in 2005 in *The National Medical Journal of India*, researchers from India selected 14 subjects to consume either rice bran oil or refined sunflower oil in their homes for three months. The nine men and five women, between the ages of 40 and 60, all had elevated levels of cholesterol. After three months, all the subjects had a washout period of three weeks before beginning a three-month use of the alternate oil. The researchers found that the use of rice bran oil for a three-month period significantly reduced the total plasma cholesterol and triglyceride levels. "The reduction in plasma LDL-cholesterol with rice bran oil was just short of statistical significance." The researchers concluded that rice bran oil "has several advantages and health benefits, it can be used routinely as a cooking oil and has the advantage that it can be consumed by the entire family."[4]

MAY BE USEFUL FOR PEOPLE WITH TYPE 2 DIABETES

In a study published in 2009 in the *Journal of Clinical Biochemistry and Nutrition*, researchers from Taiwan investigated the use of rice bran oil on lipid metabolism and insulin resistance in 16 male Wistar rats with induced type 2 diabetes. The researchers began by dividing their rats into two groups of eight. For five weeks, one group was fed a diet that consisted of 15% rice bran oil, and the other was fed a diet that consisted of 15% soybean oil. Neither diet contained any cholesterol. The researchers learned that the rats fed the rice bran oil had higher concentrations of high-density lipoprotein (HDL or "good" cholesterol) than the rats fed the soybean oil. The rats fed rice bran oil also had a lower total cholesterol/HDL ratio than the soybean fed rats. In addition, the researchers found that the body weight, weight gain, and epididymal fat weight tended to be lower in the rice bran oil group. And, their findings indicated that rice bran oil improved insulin resistance and lipid metabolism in rats with type 2 diabetes.[5]

In a randomized, single-blind, placebo-controlled study published in (July 2012) in the *Journal of Clinical Biochemistry and Nutrition*, researchers from Taiwan investigated the effect of rice bran oil consumption on plasma lipids and insulin resistance in people with type 2 diabetes. The initial cohort consisted of 40 people with type 2 diabetes between the ages of 30 and 80. For five weeks, every day, they consumed either 18g rice bran oil or 18g soybean oil. Thirty-five subjects completed the study. The researchers learned that the daily rice bran oil supplementation "significantly decreased total serum cholesterol concentrations and tended to decrease LDL-C concentrations."[6]

In a randomized, double-blind, crossover study published in 2011 in the *British Journal of Nutrition*, researchers from New Zealand attempted to determine the effectiveness of a phytosterol-containing spread derived from rice bran oil. The initial cohort consisted of 80 people, between the ages of 30 and 65, with mildly elevated levels of cholesterol. They were divided into two groups of 40 people. For four weeks, every day, the members of Group 1 consumed 20g of rice bran oil spread or standard spread or phytosterol-enriched spread. After four weeks, they switched to the next randomized spread, and four weeks later they switched to the third type of spread. Meanwhile, for four weeks, every day, the members of Group 2 consumed either 20g rice bran oil spread and 30mL rice bran oil or 20g standard spread plus 30mL sunflower oil or 20 rice bran oil spread. As with the members of Group 1, at the end of each four-week period, they went on to the next treatment. The entire trial was completed in 12 weeks.

By the end of the study, there were 75 participants. The researchers learned that when compared to the subjects who consumed the standard spread, the subjects who ate the rice bran oil spread had significant reductions in total cholesterol, LDL, and total cholesterol/HDL. However, the reductions were not as effective as those from the phytosterol-enriched spread. As for the results from Group 2, the addition of rice bran oil to rice bran oil spread produced no differences in cholesterol levels. The researchers commented that they "have no satisfactory explanation for this." They theorized that "there may be some effects resulting from the method of cooking or unreported compliance issues that need addressing." Nevertheless, the researchers concluded that their study as well as other studies "provide significant evidence that these compounds are an effective and safe adjunct to diet-controlled therapy, which can produce clinically significant reductions in LDL-cholesterol in people with mild to moderate hypercholesterolaemia."[7]

IS RICE BRAN OIL BENEFICIAL?

It is evident that rice bran oil may well have a number of beneficial properties. Most people should probably include some rice bran oil in their diets.

NOTES

1. Rice Bran Oil Website. www.ricebranoil.info.

2. Malve, H., P. Kerkar, N. Mishra et al. November 2010. "LDL-Cholesterol Lowering Activity of a Blend of Rice Bran Oil and Safflower Oil (8:2) in Patients with Hyperlipidaemia: A Proof of Concept, Double Blind, Controlled, Randomised Parallel Group Study." *Journal of the Indian Medical Association* 108(11): 785–788.

3. Utarwuthipong, T., S. Komindr, V. Pakpeankitvatana et al. January–February 2009. "Small Dense Low-Density Lipoprotein Concentration and Oxidative Susceptibility Changes after Consumption of Soybean Oil, Rice Bran Oil, Pal Oil and Mixed Rice Bran/Palm Oil in Hypercholesterolaemic Women." *The Journal of International Medical Research* 37(1): 96–104.

4. Kuriyan, R., N. Gopinath, M. Vaz, and A. V. Kurpad. November–December 2005. "Use of Rice Bran Oil in Patients with Hyperlipidaemia." *The National Medical Journal of India* 18(6): 292–296.

5. Chou, T. W., C. Y. Ma, H. H. Cheng et al. July 2009. "A Rice Bran Oil Diet Improves Lipid Abnormalities and Suppress Hyperinsulinemic Responses in Rats with Streptozotocin/Nicotinamide-Induced Type 2 Diabetes." *Journal of Clinical Biochemistry and Nutrition* 45(1): 29–36.

6. Lai, Ming-Hoang, Yi-Ting Chen, Ya-Yen Chen et al. (July 2012). "Effects of Rice Bran Oil on the Blood Lipids Profiles and Insulin Resistance in Type 2 Diabetes Patients." *Journal of Clinical Biochemistry and Nutrition* 51(1): 15–18.

7. Eady, Sarah, Alison Wallace, Jinny Willis et al. 2011. "Consumption of a Plant Sterol-Based Spread Derived from Rice Bran Oil Is Effective at Reducing Plasma Lipid Levels in Mildly Hypercholesterolaemic Individuals." *British Journal of Nutrition* 105(12): 1808–1818.

REFERENCES AND RESOURCES

Magazines, Journals, and Newspapers

Chou, T. W., C. Y. Ma, H. H, Cheng et al. July 2009. "A Rice Bran Oil Diet Improves Lipid Abnormalities and Suppress Hyperinsulinemic Responses in Rats with Streptozotocin/Nicotinamide-Induced Type 2 Diabetes." *Journal of Clinical Biochemistry and Nutrition* 45(1): 29–36.

Eady, Sarah, Alison Wallace, Jinny Willis et al. 2011. "Consumption of a Plant Sterol-Based Spread Derived from Rice Bran Oil Is Effective at Reducing Plasma Lipid Levels in Mildly Hypercholesterolaemic Individuals." *British Journal of Nutrition* 105(12): 1808–1818.

Kuriyan, R., N. Gopinath, M. Vaz, and A. V. Kurpad. November–December 2005. "Use of Rice Bran Oil in Patients with Hyperlipidaemia." *The National Medical Journal of India* 18(6): 292–296.

Lai, Ming-Hoang, Yi-Ting Chen, Ya-Yen Chen et al. (July 2012). "Effects of Rice Bran Oil on the Blood Lipids Profiles and Insulin Resistance in Type 2 Diabetes Patients." *Journal of Clinical Biochemistry and Nutrition* 51(1): 15–18.

Malve, H., P. Kerkar, N. Mishra et al. November 2010. "LDL-Cholesterol Lowering Activity of a Blend of Rice Bran Oil and Safflower Oil (8:2) in Patients with Hyperlipidaemia: A Proof of Concept, Double Blind, Controlled, Randomised Parallel Group Study." *Journal of the Indian Medical Association* 108(11): 785–788.

Utarwuthipong, T., S. Komindr, V. Pakpeankitvatana et al. January–February 2009. "Small Dense Low-Density Lipoprotein Concentration and Oxidative Susceptibility Changes after Consumption of Soybean Oil, Rice Bran Oil, Palm Oil and Mixed Rice Bran/Palm Oil in Hypercholesterolaemic Women." *The Journal of International Medical Research* 37(1): 96–104.

Websites

California Rice Oil Company. www.californiariceoil.com.
Rice Bran Oil. www.ricebraanoil.info.

Safflower Oil

Safflower seeds have truly ancient roots. They were found in the tomb of the Egyptian pharaoh Tutankhamun and are mentioned in ancient Greek writings. However, today the seeds are often pressed to make safflower oil.

Safflower oil is colorless and odorless. In addition to being used in cooking, it may be found in salad dressing, cosmetics, medicines, and commercial products.

Arthur Weisker inspects a safflower plant
in a field near Woodland, California.
(AP Photo/Rich Pedroncelli)

Some people take safflower oil supplements. It is sold in supermarkets and health food stores and is easily found online. It is generally inexpensive.

Safflower oil has high amounts of omega-3 and omega-6 fatty acids, especially linoleic acid, and good amounts of vitamin E. It is believed to support cardiovascular health, facilitate weight loss, and reduce the risk of heart disease, type 2 diabetes, and cancer.

MAY HELP OBESE PEOPLE WITH TYPE 2 DIABETES

In a 36-week, randomized, double-masked, crossover study published in 2009 in *The American Journal of Clinical Nutrition*, researchers from Ohio and Germany wanted to learn if safflower oil could be a helpful addition for overweight people with type 2 diabetes. The initial cohort consisted of 55 obese postmenopausal women with type 2 diabetes. For 16 weeks, they were placed on either conjugated linoleic acid supplementation or safflower oil supplementation. After a four-week washout period, they were then placed on the other supplementation. Thirty-five women completed the study. Although the safflower oil appeared to have no effect on the total adipose mass, "it significantly reduced trunk adipose mass and increased lean tissue mass." It also "significantly decreased fasting glucose." This is important, the researchers commented, because postmenopausal women have a greater risk for developing insulin resistance, also known as metabolic syndrome, which is characterized by extra weight around the waist, high blood pressure, and elevated levels of cholesterol.[1]

Two years later, in 2011, some of the same researchers from Ohio had a follow-up article published in *Clinical Nutrition*. In this article, they reviewed some information from the previously noted study, and they concluded with a recommendation to include safflower oil in the daily diet. They noted that "8g of SAF [safflower oil] improved glycemia, inflammation, and blood lipids, indicating that small changes in dietary fat quality may augment diabetes treatments to improve risk factors for diabetes-related complications."[2]

MAY IMPROVE BONE HEALTH

In a study published in 2010 in *Applied Physiology, Nutrition, and Metabolism*, researchers from Canada investigated the effects different high-fat diets have on femur bone mineral density, strength, and fatty acid composition. (The femur or thighbone is the largest bone in the body.) Sprague-Dawley rats, which were all 40 days old, were divided into four groups. The rats in the control group were fed a normal rat diet. For 65 days, the other rats were fed diets high in safflower oil, flaxseed oil, or coconut oil.

During the course of the study, the rats in the control group gained the least amount of weight. And, the femurs of the rats in the groups consuming safflower oil or flaxseed oil were the strongest. The researchers concluded that "the high-fat diets, containing high levels of PUFA [polyunsaturated fatty acids] in the form of flaxseed or safflower oil, have a positive effect on bone strength when fed to male rats 6 to 15 weeks of age."[3]

ANTIAGING PROPERTIES

In a study published in 2009 in *The Journals of Gerontology Series A: Biological Sciences and Medical Sciences*, researchers from Japan examined the antiaging properties of safflower oil. For 26 weeks, mice bred to age rapidly were fed diets containing 4% of one of four oils—safflower, olive, perilla, or fish. Another group of mice were fed a regular diet; that group served as the control. The researchers found that the mice fed safflower or olive oil "had fewer age-related disorders." Moreover, high-density lipoprotein (HDL or "good" cholesterol) "was significantly lower in the fish oil and perilla oil groups than in the olive oil or safflower diet group." Why is that important? The researchers explained that higher levels of HDL have been associated with "a lower risk for mortality in older men" and a "decreased risk for Alzheimer's disease."[4]

SUPPORTS CARDIOVASCULAR HEALTH

In a double-blind, controlled, parallel group study published in 2010 in the *Journal of the Indian Medical Association*, researchers based in India wanted to learn if a combination of rice bran and safflower oil would help people lower their elevated levels of cholesterol. The cohort consisted of 73 people who were randomly assigned to take the rice bran/safflower blended oil or the oil that the person normally used. Lipid profiles were monitored monthly. The researchers noted that "at each follow-up, LDL-C [low-density lipoprotein cholesterol or "bad" cholesterol] levels showed a significant reduction from baseline in the study oil group and reduction was more than that observed in the control group." By the end of the three-month study, 82% of the people who consumed the blended oil had LDL levels of less than 150mg/dL compared to 57% in the control group. The researchers commented that "the substitution of usual cooking oil with a blend of rice brand oil and safflower oil (8:2) was found to exert beneficial effects on the LDL-C levels shifting them to low-risk lipid category."[5]

MAY ENHANCE ENDURANCE

In a study published in 2011 in the *Annals of Nutrition & Metabolism*, researchers from Japan compared the effect of safflower oil, fish oil, and lard (pig fat) on the swimming endurance of aged mice. The researchers began by dividing 40 mice, which were all 58 weeks old, into three groups. One group consumed a diet that was 6% safflower oil, a second group consumed a diet that was 6% fish oil, and the third group consumed a diet that contained 6% lard. The study continued for 12 weeks. The results were noteworthy. "Mice fed safflower oil had significantly longer swimming times than mice fed lard at the end of the feeding trial, although no significant differences were observed between the two groups at the start of the study." Interestingly, the safflower oil supplementation did not have any significantly effect on body or adipose tissue weight. The researchers concluded that "safflower oil improved the swimming endurance of aged mice to a greater extent than lard."[6]

MAY BE USEFUL FOR MASSAGING NEWBORNS

In a study published in 2005 in *Indian Pediatrics*, researchers based in India compared the effects of massaging newborns with safflower oil, coconut oil, and no oil. The cohort, which consisted of 120 newborns, was divided into three groups of 40. For five days, the newborns were massaged four times a day, for 10 minutes, with one of the oils or no oil. The researchers collected pre- and postmassage blood samples and checked them for triglycerides and essential fatty acids. Interestingly, postmassage triglyceride levels rose following both types of oil massage, and also following the massages that used no oil. However, the increase was higher in the oil groups. The babies in the safflower oil massage group also had increases in essential fatty acids. The researchers concluded that "topically applied oil can be absorbed in neonates and is probably available for nutritional purposes." Moreover, "the fatty acid constituents of the oil can influence the changes in the fatty acid profiles of the massaged babies."[7]

IS SAFFLOWER OIL BENEFICIAL?

Safflower oil appears to have a number of beneficial properties. People should consider using it for their everyday cooking and other purposes.

NOTES

1. Norris, Leigh E., Angela I. Collene, Michelle L. Asp et al. September 2009. "Comparison of Dietary Conjugated Linoleic Acid with Safflower Oil on Body Composition in Obese Postmenopausal Women with Type 2 Diabetes Mellitus." *The American Journal of Clinical Nutrition* 90(3): 468–476.

2. Asp, Michelle L., Angela L. Collene, Leigh E. Norris et al. 2011. "Time-Dependent Effects of Safflower Oil to Improve Glycemia, Inflammation and Blood Lipids in Obese, Post-Menopausal Women with Type 2 Diabetes: A Randomized, Double-Masked, Cross-over Study." *Clinical Nutrition* 30(4): 443–449.

3. Lau, Beatrice Y., Val Andrew Fajardo, Lauren McMeekin et al. October 2010. "Influence of High-Fat Diet from Differential Dietary Sources on Bone Mineral Density, Bone Strength, and Bone Fatty Acid Composition in Rats." *Applied Physiology, Nutrition, and Metabolism* 35(5): 598–606.

4. Umezawa, M., K. Higuchi, M. Mori et al. June 2009. "Effect of Dietary Unsaturated Fatty Acids on Senile Amyloidosis in Senescence-Accelerated Mice." *The Journals of Gerontology Series A: Biological Sciences and Medical Sciences* 64(6): 646–652.

5. Malve, H., P. Kerkar, N. Mishra et al. November 2010. "LDL-Cholesterol Lowering Activity of a Blend of Rice Bran Oil and Safflower Oil (8:2) in Patients with Hyperlipidaemia: A Proof of Concept, Double Blind, Controlled, Randomised Parallel Group." *Journal of the Indian Medical Association* 108(11): 785–788.

6. Zhang, G., N. Shirai, and H. Suzuki. October 2011. "Relationship between the Effect of Dietary Fat on Swimming Endurance and Energy Metabolism in Aged Mice." *Annals of Nutrition & Metabolism* 58(4): 282–289.

7. Solanki, K., M. Matnani, M. Kale et al. October 2005. "Transcutaneous Absorption of Topically Massaged Oil in Neonates." *Indian Pediatrics* 42(10): 998–1005.

REFERENCES AND RESOURCES

Magazines, Journals, and Newspapers

Asp, Michelle L., Angela L. Collene, Leigh E. Norris et al. 2011. "Time-Dependent Effects of Safflower Oil to Improve Glycemia, Inflammation, and Blood Lipids in Obese, Post-Menopausal Women with Type 2 Diabetes: A Randomized, Double-Masked, Crossover Study." *Clinical Nutrition* 30(4): 443–449.

Baumann, Leslie S. August 2012. "Safflower Oil." *Skin & Allergy News* 43(8): 26.

Lau, Beatrice Y., Val Andrew Fajardo, Lauren McMeekin et al. October 2010. "Influence of High-Fat Diet from Differential Dietary Sources on Bone Mineral Density, Bone Strength, and Bone Fatty Acid Composition in Rats." *Applied Physiology, Nutrition, and Metabolism* 35(5): 598–606.

Malve, H., P. Kerkar, N. Mishra et al. November 2010. "LDL-Cholesterol Lowering Activity of a Blend of Rice Bran Oil and Safflower Oil (8:2) in Patients with Hyperlipidaemia: A Proof of Concept, Double Blind, Controlled, Randomised Parallel Group Study." *Journal of the Indian Medical Association* 108(11): 785–788.

Norris, Leigh E., Angela L. Collene, Michelle L. Asp et al. September 2009. "Comparison of Dietary Conjugated Linoleic Acid with Safflower Oil on Body Composition in Obese Postmenopausal Women with Type 2 Diabetes Mellitus." *The American Journal of Clinical Nutrition* 90(3): 468–476.

Solanki, K., M. Matnani, M. Kale et al. October 2005. "Transcutaneous Absorption of Topically Massaged Oil in Neonates." *Indian Pediatrics* 42(10): 998–1005.

Umezawa, M., K. Higuchi, M. Mori et al. June 2009. "Effect of Dietary Unsaturated Fatty Acids on Senile Amyloidosis in Senescence-Accelerated Mice." *The Journals of Gerontology Series A: Biological Sciences and Medical Sciences* 64A(6): 646–652.

Zhang, G., N. Shirai, and H. Suzuki. October 2011. "Relationship between the Effect of Dietary Fat on Swimming Endurance and Energy Metabolism in Aged Mice." *Annals of Nutrition & Metabolism* 58(4): 282–289.

Website

Safflower Oil. http://safflower-oil.org.

Salmon Oil: Overview

Like the oil of other types of fatty fish, salmon oil contains abundant amounts of unsaturated fats known as omega-3 essential fatty acids. Though the body is unable to make omega-3 fatty acids, they are believed to have a host of benefits, such as supporting cardiovascular health, lowering elevated blood pressure, and reducing the symptoms of some psychiatric illnesses, including depression. Omega-3 fatty acids may also ease the joint pain associated with arthritis and help people lose weight. There are three main types of these fatty acids—alpha-linolenic acid (ALA), eicosapentaenoic acid (EPA), and docosahexaenoic acid (DHA).[1]

Salmon oil is readily available in supermarkets and specialty stores. It may also be easily purchased online. Whenever possible, it is best to purchase salmon oil that has been produced from wild salmon, as opposed to farm-raised salmon. Salmon oil made from wild salmon may cost a little more, but it will probably be a healthier supplement.

This entry discusses the association between salmon oil and a variety of medical concerns. The next entry addresses the association between salmon oil and memory function.

MAY HELP CHILDREN WITH ATTENTION-DEFICIT/ HYPERACTIVITY DISORDER

In a randomized, controlled study published in 2012 in *Nutrition*, researchers from Australia attempted to determine if a supplementation with EPA or DHA would help improve the literacy and behavior problems associated with children with attention-deficit/hyperactivity disorder (ADHD). The initial cohort consisted of 90 children between the ages of 7 and 12 who were diagnosed with ADHD. For four months, the children took supplements containing EPA, DHA, or safflower oil (control). Seventy children completed the trial. The researchers found that the children with higher levels of DHA in their red blood cells had improved reading and spelling. In addition, their parents reported that their ADHD symptoms had somewhat moderated. The researchers noted that their findings "support indications that children with learning difficulties as part of a constellation of developmental problems may be a subgroup of responders to ω-3 PUFA [omega-3 polyunsaturated fatty acid] supplementation."[2]

MAY HELP REDUCE THE BRAIN'S RESPONSE TO A DIET HIGH IN FRUCTOSE

In a study published in 2012 in *The Journal of Physiology*, researchers from Los Angeles, California (United States), wanted to learn if DHA could be of any assistance to the brain of rats that have been compromised by a high-fructose (type of sugar) diet. For six weeks, Sprague-Dawley rats were fed a high-fructose solution in drinking water. One group of rats also received DHA. The researchers found that a high-fructose diet hindered memory and learning. However, the addition of DHA to the diet minimized the damage and improved memory. They noted that "it is encouraging that the presence of the n-3 [DHA] diet was sufficient to buffer the effects of metabolic dysfunction." And, they concluded that "in terms of public health, these results support the encouraging possibility that health diets can attenuate the action of unhealthy diets such that the right combination of foods is crucial for a health brain."[3]

MAY BE USEFUL FOR THE SIDE EFFECTS OF MEDICATION TREATMENTS FOR HIV

In an open label, randomized, parallel, multicenter, crossover study published in 2007 in *HIV Clinical Trials*, researchers from Canada noted that the treatment for HIV (human immunodeficiency virus) is called HAART or highly active antiretroviral therapy. While it has proven to be quite useful for HIV, one of its side effects is the elevation of serum triglycerides—fats in the blood that provide energy to the body. And, elevated levels of serum triglycerides are associated with cardiovascular disease. The participants in the study took 1g salmon oil three times per day for 24 weeks or no treatment for 12 weeks and salmon oil treatment for weeks 12–24. Fifty-eight participants completed the study. The researchers found that the addition of salmon oil to the diet resulted in significant reductions in the levels of triglycerides. And, they concluded that "low-dose salmon oil (3g/day) is effective and well-tolerated in reducing TG [triglyceride] levels in HIV-infected patients receiving HAART."[4]

IMPROVES THE AMOUNT OF LONG-CHAIN OMEGA-3 FATTY ACIDS IN THE BLOOD

In a study published in 2011 in the *British Journal of Nutrition*, researchers from New Zealand compared the effects of the consumption of farmed New Zealand King Salmon and the daily intake of salmon oil capsules on the levels of long-chain omega-3 fatty acids and selenium in the blood. According to the researchers, these results are particularly important to people who reside in New Zealand, who have "marginal" levels of these nutrients. The initial cohort consisted of 44 healthy volunteers between the ages of 21 and 40. They were randomly assigned to one of four groups. In the first group, the volunteers ate 120g servings of salmon twice each week for eight weeks. In the second, third, and fourth groups, the volunteers took 2, 4, or 6 salmon oil capsules per day, also for eight weeks. The researchers found that the "consumption of similar amounts of LC n-3 fatty acids ether from two weekly servings of salmon or daily dosages of salmon oil capsules was equally effective in increasing LC *n-3* status in healthy volunteers." However, "consuming salmon had the added benefit of increasing plasma Se [selenium] concentrations and was better tolerated than salmon oil capsules."[5]

MAY BE USEFUL IN PREVENTING COLORECTAL CANCER

In a study published in 2007 in the *American Journal of Epidemiology*, researchers from the United Kingdom wanted to learn if fatty acids could play a role in the prevention of colorectal cancer. The cohort consisted of 1,455 people with colorectal cancer and 1,455 matched controls. More than 99% of the study participants were white (Caucasian). When the researchers compared the people with colorectal cancer to the controls, they found that the people with cancer ate lower amounts of omega-3 polyunsaturated fatty acids, specifically EPA and DHA. And, "the effects remained constant and significant after further energy adjustment and stratification."[6]

MAY HELP PREVENT PERIODONTAL DISEASE

In a study published in 2010 in *Nutrition*, researchers from Japan, the United Kingdom, and Michigan (United States) investigated the association between intake of omega-3 fatty acids, primarily DHA and EPA, and periodontal disease (inflammation of the gums and loss of attachment of the periodontal ligament) in the elderly. Why is this important? Periodontal disease is the primary reason for tooth loss among this age group. The cohort consisted of 55 volunteers who were all 74 years. Dental examinations were conducted at the beginning of the study and once each year for five years. During the course of the study, 19 participants withdrew. The researchers found that the participants with the lowest intake of DHA had the fewest numbers of teeth. These results were statistically significant. Though the resulted obtained for EPA were somewhat similar, they were not statistically significant. The researchers concluded that there may be an inverse association between intake of DHA and the progression of periodontal disease in older people.[7]

MAY HELP WITH DEPRESSIVE SYMPTOMS

In a study published in 2009 in *Nutrition*, researchers from Chicago, Illinois; Chapel Hill, North Carolina; and San Francisco, California (all three United States), wanted to determine if there is an association between the intake of omega-3 fatty acids and the incidence of depressive symptoms in women. Employing data from the Coronary Artery Risk Development in Young Adults study, the researchers found that for the entire cohort, the highest intake of DHA, EPA, and DHA plus EPA were associated with the decreased risk for the symptoms of depression. Moreover, "the observed inverse associations were more pronounced for women."[8]

In a randomized, masked, placebo-controlled study published in 2012 in the *Journal of Clinical Psychopharmacology*, researchers from California and Tennessee (both United States) compared the use of citalopram, a selective serotonin reuptake inhibitor medication for depression, combined with omega-3 fatty acids to the use of citalopram plus olive oil (placebo) for treating depression. The initial cohort consisted of 42 subjects. For nine weeks, the subjects took either a combination therapy (two 1g capsules containing a blend of 900mg EPA, 200mg of DHA, and 100mg of other omega-3 fatty acids, twice daily plus citalopram) or monotherapy (two 1g capsules of olive oil per day plus citalopram). The researchers found that the participants on the combination therapy experienced significant improvements in their depressive symptoms. "Our data suggest that initiation of treatment with a selective serotonin reuptake inhibitor and PUFA simultaneously is advantageous in efficacy when compared with treatment with selective serotonin update inhibitor as a monotherapy. The safety profile of such as combination treatment is very favorable."[9]

IS SALMON OIL BENEFICIAL?

Though there are mixed reports on the benefits of salmon oil, it clearly may be useful for some people with certain medical concerns. Before beginning a regime of salmon oil, it is important to have a discussion with a medical provider.

NOTES

1. LIVESTRONG.COM. http://livestrong.com.

2. Milte, Catherine M., Natalie Parletta, Jonathan D. Buckley et al. 2012. "Eicosapentaenoic and Docosahexaenoic Acids, Cognition, and Behavior in Children with Attention-Deficit/Hyperactivity Disorder: A Randomized Controlled Trial." *Nutrition* 28(6): 670–677.

3. Agrawal, R. and F. Gomez-Pinilla. May 15, 2012. "'Metabolic Syndrome' in the Brain: Deficiency in Omega-3 Fatty Acid Exacerbates Dysfunctions in Insulin Receptor Signaling and Cognition." *The Journal of Physiology* 590(10): 2485–2499.

4. Baril, Jean-Guy, Colin M. Kovacs, Sylvie Trottier et al. November–December 2007. "Effectiveness and Tolerability of Oral Administration of Low-Dose Salmon Oil to HIV Patients with HAART Associated Dyslipidemia." *HIV Clinical Trials* 8(6): 400–411.

5. Stonehouse, W., M.R. Pauga, R. Kruger et al. October 28, 2011. "Consumption of Salmon v. Salmon Oil Capsules: Effects on N-3 PUFA and Selenium Status." *British Journal of Nutrition* 106(8): 1231–1239.

6. Theodoratou, Evropi, Geraldine McNeill, Roseanne Cetnarskyj et al. 2007. "Dietary Fatty Acids and Colorectal Cancer: A Case-Control Study." *American Journal of Epidemiology* 166(2): 181–195.

7. Iwasaki, Masanori, Akihiro Yoshihara, Paula Moynihan et al. 2010. "Longitudinal Relationship between Dietary ω-3 Fatty Acids and Periodontal Disease." *Nutrition* 26(11–12): 1105–1109.

8. Colangelo, Laura A., Ka He, Mary A. Whooley et al. 2009. "Higher Dietary Intake of Long-Chain ω-3 Polyunsaturated Fatty Acids Is Inversely Associated with Depressive Symptoms in Women." *Nutrition* 25(10): 1011–1019.

9. Gertsik, L., R.E. Poland, C. Bresee, and M.H. Rapaport. February 2012. "Omega-3 Fatty Acid Augmentation of Citalopram Treatment for Patients with Major Depressive Disorder." *Journal of Clinical Psychopharmacology* 32(1): 61–64.

REFERENCES AND RESOURCES

Magazines, Journals, and Newspapers

Agrawal, R. and F. Gomez-Pinilla. May 15, 2012. "'Metabolic Syndrome' in the Brain: Deficiency in Omega-3 Fatty Acid Exacerbates Dysfunctions in Insulin Receptor Signaling and Cognition." *The Journal of Physiology* 590(10): 2485–2499.

Baril, Jean-Guy, Colin M. Kovacs, Sylvie Trottier et al. November–December 2007. "Effectiveness and Tolerability of Oral Administration of Low-Dose Salmon Oil to HIV Patients with HAART Associated Dyslipidemia." *HIV Clinical Trials* 8(6): 400–411.

Colangelo, Laura A., Ka He, Mary A. Whooley et al. 2009. "Higher Dietary Intake of Long-Chain ω-3 Polyunsaturated Fatty Acids Is Inversely Associated with Depressive Symptoms in Women." *Nutrition* 25(10): 1011–1019.

Gertsik, L., R.E. Poland, C. Bresee, and M.H. Rapaport. February 2012. "Omega-3 Fatty Acid Augmentation of Citalopram Treatment for Patients with Major Depressive Disorder." *Journal of Clinical Psychopharmacology* 32(1): 61–64.

Iwasaki, Masanori, Akihiro Yoshihara, Paula Moynihan et al. 2010. "Longitudinal Relationship between Dietary ω-3 Fatty Acids and Periodontal Disease." *Nutrition* 26(11–12): 1105–1109.

Milte, Catherine M., Natalie Parletta, Jonathan D. Buckley et al. 2012. "Eicosapentaenoic and Docosahexaenoic Acids, Cognition, and Behavior in Children with Attention-Deficit/Hyperactivity Disorder: A Randomized Controlled Trial." *Nutrition* 28(6): 670–677.

Stonehouse, W., M.R. Pauga, R. Kruger et al. October 28, 2011. "Consumption of Salmon v. Salmon Oil Capsules: Effects on N-3 PUFA and Selenium Status." *British Journal of Nutrition* 106(8): 1231–1239.

Theodoratou, Evropi, Geraldine McNeill, Roseanne Cetnarskyj et al. 2007. "Dietary Fatty Acids and Colorectal Cancer: A Case Control Study." *American Journal of Epidemiology* 166(2): 181–195.

Websites

LIVESTRONG.COM. http://livestrong.com.

University of Maryland Medical Center. www.umm.edu.

Salmon Oil and Memory Function

In the previous entry, we examined studies that explored the usefulness of salmon oil for a myriad of health benefits. In this entry, we discuss the studies that both support and repute the use of salmon oil for improving memory.

MAY HELP WITH MEMORY PROBLEMS

In a study published in 2012 in *Neurology*, researchers from several locations in the United States examined the levels of docosahexaenoic acid(DHA) and eicosapentaenoic acid(EPA) in the red blood cells of more than 1,500 adults without dementia. Their average age was 67. The researchers found an association between lower levels of DHA and "markers of accelerated structural and cognitive aging." People with DHA levels that were so low that they were in the 25 percentile had substantially lower brain volume that those with higher levels of DHA. When the researchers conducted magnetic resonance imaging on the participants, they determined that differences between the people with low and higher levels of DHA was equivalent to about two years of structural brain aging. Moreover, participants who had the lowest amounts of overall omega-3 levels had the worse results in testing that measured visual memory, abstract thinking, and executive function. These include organizing, planning, and recalling information.[1]

In a randomized, double-blind, placebo-controlled study published in 2010 in *Alzheimer's & Dementia: The Journal of the Alzheimer's Association*, researchers from Maryland, Pennsylvania, Florida, and the United Kingdom examined the association between the consumption of DHA and memory in older adults with mild cognitive impairments (MCIs). The cohort consisted of 485 subjects; everyone was 55 or older and voiced complaints about their memory. For 24 weeks, every day, subjects took either 900mg of DHA or a placebo. The researchers found that the DHA improved memory and learning in this group of adults. "Twenty-four week supplementation with 900 mg/d DHA improved learning and memory function in ARCD [age-related cognitive decline] and is a beneficial supplement that supports cognitive health with aging."[2]

In a study published in 2012 in *Neurology*, researchers based at Columbia University in New York City investigated the association between intake of omega-3 fatty acids and Alzheimer's disease. The cohort consisted of 1,219 people 65 and older who did not have dementia. For more than a year, the researchers tracked their intake of food and nutrients, including omega-3 fatty acids. At the end of the trial, they tested blood levels of beta-amyloid. The researchers found that higher levels of omega-3 were related to lower levels of beta-amyloid. Why is that important? Beta-amyloid has been linked to Alzheimer's disease. So, the consumption of omega-3 fatty acid, such as salmon oil, suggests "the potential beneficial effects of ω-3 PUFA [omega-3 polyunsaturated fatty acids] on AD [Alzheimer's disease] and cognitive function."[3]

BUT THE RESULTS VARY MARKEDLY

In a 24-week, randomized, double-blind, placebo-controlled study published in 2008 in *Progress in Neuro-Psychopharmacology & Biological Psychiatry*, researchers from Taiwan tested the ability of omega-3 supplementation to help people with less advanced Alzheimer's disease. Every day, 23 participants with mild to moderate Alzheimer's disease and 23 participants with MCI received either 1.8g omega-3 or a placebo (olive oil). The data of 35 (76%) of the participants were-analyzed. The researchers found that "omega-3 fatty acids may improve general clinical function in patients with mild or moderate MCI, but not their cognitive function." In addition, "the cognitive effects of omega-3 fatty acids might be favored in patients with MCI rather than those with AD."[4]

In a double-blind, controlled trial published in 2010 in *The American Journal of Clinical Nutrition*, researchers from the United Kingdom and Australia wanted to learn if omega-3 polyunsaturated fatty acid supplementation would benefit cognitive function in healthy older adults. For 24 months, a total of 867 adults between the ages of 70 and 79 were randomly assigned to take daily capsules providing 200mg EPA and 500mg DHA or olive oil. During the course of the trial, "cognitive function did not decline in either study arm." The researchers concluded that the supplementation did not appear to benefit their cognitively healthy older people. But, there was also no evidence of any harm.[5]

In a randomized, double-blind, placebo-controlled study published in 2010 in *JAMA: The Journal of the American Medical Association*, researchers from several locations in the United States attempted to determine if supplementation with DHA would slow the cognitive and functional decline seen in people with mild to moderate Alzheimer's disease. The participants in the initial cohort, which contained 402 people, were assigned to take either a daily dose of 2g DHA or an identical placebo. At the end of 18 months, a total of 295 people completed the study (DHA: 171, placebo: 124). The researchers found that the DHA supplementation "did not slow the rate of cognitive and functional decline in patients with mild to moderate Alzheimer's disease." And, they concluded that DHA was not useful for this population.[6]

Over a period of six months, Welma Stonehouse, PhD, and her fellow researchers from Massey University in New Zealand gave supplements containing DHA to 176 healthy adults. During this time, they evaluated memory and cognitive functioning and compared the results to adults taking placebos. The researchers found that memory, working memory, and the speed of working memory in the adults taking DHA showed significant improvement. According to Dr. Stonehouse, "this is the first robust study to show that a DHA-rich supplement can improve some aspects of memory functioning in young healthy adults." She added that "the cognitive functions shown to be affected by the DHA-rich fish oil, namely memory and working memory, are among the most important functions of our brains for numerous everyday activities, such as working, driving, shopping studying, playing sports etc." She stressed that "maintaining brain health and getting your brain to perform at its optimal capacity is just as vital as maintaining physical well-being and health."[7]

In a review published in 2012 in the *Cochrane Database of Systematic Reviews*, researchers from the United Kingdome and Singapore evaluated the findings of randomized, controlled studies that attempted to determine if the higher consumption of omega-3 protects cognitively healthy people over the age of 60 from cognitive decline and dementia. After examining the studies, the researchers found that omega-3 provided no memory benefit to cognitively healthy older people. However, they do acknowledge that omega-3 supplementation may support health in another way. The researchers concluded that "direct evidence of the effect of omega-3 PUFA on incident dementia is lacking."[8]

IS SALMON OIL USEFUL FOR MEMORY PROBLEMS?

Obviously, there is no definitive answer to that question. Nevertheless, since salmon oil is so reasonably priced, people who are dealing with memory problems may wish to give it a try, if only for a few months.

NOTES

1. Tan, Z.S., W.S. Harris, A.S. Beiser et al. February 28, 2012. "Red Blood Cell Omega-3 Fatty Acid Levels and Markers of Accelerated Brain Aging." *Neurology* 78(9): 658–664.

2. Yurko-Mauro, Karin, Deanna McCarthy, Dror Rom et al. November 2010. "Beneficial Effects of Docosahexaenoic Acid on Cognition in Age-Related Cognitive Decline." *Alzheimer's & Dementia: The Journal of the Alzheimer's Association* 6(6): 456–464.

3. Gu, Y., N. Schupf, S.A. Cosentino et al. June 5, 2012. "Nutrient Intake and Plasma β-Amyloid." *Neurology* 78(23): 1832–1840.

4. Chiu, Chih-Chiabg, Kuan-Pin Su, Tsung-Chi Cheng et al. 2008. "The Effects of Omega-3 Fatty Acids Monotherapy in Alzheimer's Disease and Mild Cognitive Impairment: A Preliminary Randomized Double-Blind Placebo-Controlled Study." *Progress in Neuro-Psychopharmacology & Biological Psychiatry* 32(6): 1538–1544.

5. Dangour, Alan D., Elizabeth Allen, Diana Elbourne et al. June 2010. "Effect of 2-Y N-3 Long-Chain Polyunsaturated Fatty Acid Supplementation on Cognitive Function in Older People: A Randomized, Double-Blind, Controlled Trial." *The American Journal of Clinical Nutrition* 91(6): 1725–1732.

6. Quinn, Joseph F., Rema Raman, Ronald G. Thomas et al. November 3, 2010. "Docosahexaenoic Acid Supplementation and Cognitive Decline in Alzheimer Disease." *JAMA: The Journal of the American Medical Association* 304(17): 1903–1911.

7. Daily Mail Website. http://www.dailymail.co.uk.

8. Sydenham, Emma, Alan D. Dangour, and Wee-Shiong Lim. 2012. "Omega 3 Fatty Acid for the Prevention of Cognitive Decline and Dementia." *Cochrane Database of Systematic Reviews* Issue 6 Article Number CD005379.

REFERENCES AND RESOURCES
Magazines, Journals, and Newspapers

Chiu, Chih-Chiang, Kuan-Pin Su, Tsung-Chi Cheng et al. 2008. "The Effects of Omega-3 Fatty Acids Monotherapy in Alzheimer's Disease and Mild Cognitive Impairment: A Preliminary Randomized Double-Blind Placebo-Controlled Study." *Progress in Neuro-Psychopharmacology & Biological Psychiatry* 32(6): 1538–1544.

Dangour, Alan D., Elizabeth Allen, Diana Elbourne et al. June 2010. "Effect of 2-Y N-3 Long-Chain Polyunsaturated Fatty Acid Supplementation on Cognitive Function in Older People: A Randomized, Double-Blind, Controlled Trial." *The American Journal of Clinical Nutrition* 91(6): 1725–1732.

Gu, Y., N. Schupf, S. A. Cosentino et al. June 5, 2012. "Nutrient Intake and Plasma β-Amyloid." *Neurology* 78(23): 1832–1840.

Quinn, Joseph F., Rema Ramen, Ronald G. Thomas et al. November 3, 2010. "Docosahexaenoic Acid Supplementation and Cognitive Decline in Alzheimer Disease: A Randomized Trial." *JAMA: The Journal of the American Medical Association* 304(17): 1903–1911.

Sydenham, Emma, Alan D. Dangour, and Wee-Shiong Lim. 2012. "Omega 3 Fatty Acid for the Prevention of Cognitive Decline and Dementia." *Cochrane Database of Systematic Reviews* Issue 6 Article Number CD005379.

Tan, Z. S., W. S. Harris, A. S. Beiser et al. February 28, 2012. "Red Blood Cell Omega-3 Fatty Acid Levels and Markers of Accelerated Brain Aging." *Neurology* 78(9): 658–664.

Yurko-Mauro, Karin, Deanna McCarthy, Dror Rom et al. November 2010. "Benificial Effects of Docosahexaenoic Acid on Cognition in Age-Related Cognitive Decline." *Alzheimer's & Dementia: The Journal of the Alzheimer's Association* 6(6): 456–464.

Websites

Cochrane Summaries. http://summaries.cochrane.org.
Daily Mail. http://www.dailymail.co.uk.
Kaumudi Global. www.kaumudiglobal.com.
LIVESTRONG.COM. http://livestrong.com.

Saw Palmetto Oil

For years, saw palmetto oil and other products made from the saw palmetto plant have been marketed for two main medical problems. Saw palmetto is probably best known as a treatment for benign prostatic hyperplasia (BPH) or the enlargement of the prostate gland in men. In fact, a 2011 article in *Better Nutrition* magazine was entitled "The 'Must-Have' Herb for the Men over 40."[1] But, it is also thought to be an effective treatment for urinary tract problems. To a lesser extent, it has been said to be useful for baldness, colds and coughs, sore throat, migraines, bronchitis, and asthma.[2]

Saw palmetto oil should be available in large discount and drug stores. If not, they are easily purchased online. It is moderately priced.

MAY OR MAY NOT BE EFFECTIVE TREATMENT
FOR BENIGN PROSTATIC HYPERPLASIA

In a double-blind, placebo-controlled study published in 2008 in *The Journal of Urology*, researchers from China and Los Angeles, California, investigated the ability of saw palmetto to help those dealing with lower urinary tract symptoms associated with an enlarged prostate gland. The initial cohort consisted of 94 Chinese men between the ages of 49 and 75 who were newly diagnosed with BPH. They had a variety of symptoms such as nocturia (getting up during the night to urinate), incomplete emptying, urinary frequency, straining, a weak urine stream, and urgency. For 12 weeks, 46 of the men took two daily soft gels in which the primary ingredient was saw palmetto and 48 of the men took two placebo soft gels daily. Ninety-two men completed the trial. While the researchers found that the soft gels improved some of the symptoms related to the enlarged prostates, the soft gels "had no significant effect on prostate size." And, they concluded that saw palmetto oil "may have short-term effects in improving symptoms and objective measures in Chinese men with lower urinary tract symptoms associated with benign prostatic hyperplasia."[3]

In a study published in 2011 in *The Journal of Urology*, researchers from Italy compared the effects of treating rats with saw palmetto versus treating them with saw palmetto, selenium, and lycopene. The cohort consisted of 28 Sprague-Dawley rats that were randomly assigned to one of four groups. One group served as the control. The remaining three groups of rats were all treated with testosterone, to promote growth of the prostate gland. One group only received the testosterone. A second group received the testosterone and saw palmetto supplementation; and the third group received testosterone, saw palmetto, selenium, and lycopene. The various treatments continued for 14 days. After the rats were sacrificed, their prostate glands were removed for analysis. As the researchers expected, the rats that received only testosterone experienced a notable enlargement of their prostate glands. The rats on testosterone and saw palmetto had less prostate enlargement; but, the rats on testosterone and saw palmetto, selenium, and lycopene had even better reductions in the growth of their prostate glands. The researchers concluded that the combination of saw palmetto, selenium, and lycopene is "a possible alternative therapy for hormone dependent prostatic growth."[4]

On the other hand, a randomized, double-blind, multicenter, placebo-controlled study published in 2011 in *JAMA The Journal of the American Medical Association* reported very different findings. In this trial, researchers from Canada and different locations in the United States tested the ability of as much as three times the normal dose of saw palmetto to control symptoms in men caused by enlarged prostate glands. The cohort consisted of 369 men, aged 45 or older, from 11 different North American clinical sites. For 72 weeks, the men took either saw palmetto or a placebo. Though the men on saw palmetto initially took a single dose, that dose was doubled at week 24 and tripled at week 48. Surprisingly, by the end of the trial, there was little difference between the treatment and placebo groups. The researchers concluded that saw palmetto "used up to 3 times the standard daily dose had no greater effect than placebo."[5]

TREATMENT FOR LOWER URINARY TRACT SYMPTOMS

In a study published in 2010 in *Urologia*, researchers based in Italy noted that saw palmetto has been used to treat lower urinary tract symptoms. It is believed to have "antiandrogenic [blocks the action of androgens, the hormones responsible for male characteristics], antiproliferative [inhibiting cell growth] and anti-inflammatory properties." They decided to combine saw palmetto with stinging nettle, which has antiproliferative properties, and with pine bark extract, which has antinflammatory properties. During the years 2007 and 2008, the researchers treated 320 patients suffering lower urinary tract symptoms with the three-part combination for a minimum of 30 days to a maximum of one year. The subjects ranged in age between 19 and 78. Forty-six percent of the patients had BPH, 43% had chronic prostatitis syndrome, 7% had chronic genital-pelvic pain, and 4% had other conditions. Complete evaluations could only be conducted on 80 of the 320 patients. Still, 68 of the 80 people (85%) "reported a significant benefit, with special reference to an improvement of pain, urgency, stanguary [frequent, painful urination of small amounts of urine] and nocturia." Data on prostate volume were available on 84 men. The researchers found no significant changes. They commented that the combination therapy was "safe and well tolerated."[6]

MAY HELP IN THE CONTROL OF OILY SKIN

In a study published in 2007 in the *Journal of Cosmetic Dermatology*, a researcher from Bulgaria wanted to determine if a cream containing saw palmetto, sesame seeds, and argan oil was useful in the control of excessively oily skin. The cohort consisted of 20 healthy volunteers (9 male and 11 female) between the ages of 17 and 50. Sixteen of the volunteers had oily skin; four had combination skin. The study was conducted during the winter months of January and February. For four weeks, the subjects applied the test cream to the face twice daily. All of the participants completed the study. The researchers found "a significant reduction of sebum level on the skin (20% decrease) and area covered by oily spots (42% decrease) after 4 weeks of twice-daily applications." The cream "reduced the greasiness and improved that appearance of oily facial skin."[7]

MAY BE USEFUL FOR MOST COMMON
FORM OF HAIR LOSS

In a randomized, double-blind study published in 2002 in *The Journal of Alternative and Complementary Medicine*, researchers from Aurora, Colorado; Denver, Colorado; and Atlanta, Georgia, attempted to learn if saw palmetto and beta-sitosterol, a substance found in plants, would be useful for androgenetic alopecia (AGA), the absence of hair from skin area in which it is normally present—the most common form of hair loss. The initial cohort consisted of

26 male subjects between the ages of 23 and 64; all of the subjects had mild to moderate AGA. During the study, which continued for 18–24.7 weeks, the men took a supplement with 400mg of saw palmetto and 100mg of beta-sitosterol or a placebo. Nineteen men completed the trial. The researchers found that 60% of the men taking the supplement had more hair growth than the men taking the placebo. They concluded that they "observed objective evidence of efficacy using orally administered botanical therapy in the treatment of AGA, for the first time."[8]

IS SAW PALMETTO BENEFICIAL?

An article published in 2012 in the *Journal of Family Practice* noted that a survey conducted in 2007 found that 1.6 million adults in the U.S reported using saw palmetto, usually for an enlarged prostate, in the 30 days prior to the survey. So, obviously, there are many people who believe that saw palmetto is beneficial. But, the available research is not conclusive. Still, since there are almost no reports of toxic reactions to saw palmetto, there is the possibility that some people suffering from certain medical problems may benefit.

NOTES

1. Bowdeo, Jonnv. June 2011. "The 'Must-Have' Herb for the Men over 40." *Better Nutrition* 73(6): 27.

2. MedlinePlus. www.nlm.nih.gov/medlineplus.

3. Shi, Rong, Qiungwen Xie, X. Gang et al. February 2008. "Effect of Saw Palmetto Soft Gel Capsule on Lower Urinary Tract Symptoms Associated with Benign Prostatic Hyperplasia: A Randomized Trial in Shanghai, China." *The Journal of Urology* 179(2): 610–615.

4. Altavilla, Domenica, Alessandra Bitto, Francesca Polito et al. October 2011. "The Combination of *Serenoa repens*, Selenium, and Lycopene Is More Effective Than *Serenoa repens* Alone to Prevent Hormone Dependent Prostatic Growth." *The Journal of Urology* 186(4): 1524–1529.

5. Barry, Michael J., Sreelatha Meleth, Jeannette Y. Lee et al. September 28, 2011. "Effect of Increasing Doses of Saw Palmetto Extract on Lower Urinary Tract Symptoms: A Randomized Trial." JAMA: *The Journal of the American Medical Association* 306(12): 1344–1351.

6. Pavone, C., D. Abbadessa, M.L. Tarantino et al. January–March 2010. "Associating *Serenoa repens*, *Urtica dioica*, and *Pinus pinaster*. Safety and Efficacy in the Treatment of Lower Urinary Tract Symptoms. Prospective Study on 320 Patients." *Urologia* 77(1): 43–51.

7. Dobrev, Hristo. June 2007. "Clinical and Instrumental Study of the Efficacy of a New Sebum Control Cream." *Journal of Cosmetic Dermatology* 6(2): 113–118.

8. Prager, Nelson, Karen Bickett, Nita French, and Geno Marcovici. April 2002. "A Randomized, Double-Blind, Placebo-Controlled Trial to Determine the Effectiveness of Botanically Derived Inhibitors of 5-α-Reductase in the treatment of Androgenetic Alopecia." *The Journal of Alternative and Complementary Medicine* 8(2): 143–152.

REFERENCES AND RESOURCES

Magazines, Journals, and Newspapers

Altavilla, Domenica, Alessandra Bitto, Francesca Polito et al. October 2011. "The Combination of *Serenoa repens*, Selenium and Lycopene Is More Effective Than *Serenoa repens* Alone to Prevent Hormone Dependent Prostatic Growth." *The Journal of Urology* 186(4): 1524–1529.

Barry, Michael J., Sreelatha Meleth, Jeannette Y. Lee, et al. September 28, 2011. "Effect of Increasing Doses of Saw Palmetto Extract on Lower Urinary Tract Symptoms: A Randomized Trial." *JAMA: The Journal of the American Medical Association* 306(12): 1344–1351.

Bowdeo, Jonnv. June 2011. "The 'Must-Have' Herb for the Men over 40." *Better Nutrition* 73(6): 27.

Dobrev, Hristo. June 2007. "Clinical and Instrumental Study of the Efficacy of a New Sebum Control Cream." *Journal of Cosmetic Dermatology* 6(2): 113–118.

Pavone, C., D. Abbadessa, M. L. Tarantino et al. January–March 2010. "Associating *Serenoa repens*, *Urtica dioica*, and *Pinus pinaster*. Safety and Efficacy in the Treatment of Lower Urinary Tract Symptoms. Prospective Study on 320 Patients." *Urologia* 77(1): 43–51.

Prager, Nelson, Karen Bickett, Nita French, and Geno Marcovici. April 2002. "A Randomized, Double-Blind, Placebo-Controlled Trial to Determine the Effectiveness of Botanically Derived Inhibitors of 5-α-Reductase in the Treatment of Androgenetic Alopecia." *The Journal of Alternative and Complementary Medicine* 8(2): 143–152.

Ricco, Jason and Shailendra Prasad. July 2012. "The Shrinking Case for Saw Palmetto: Findings of This Study Should Make Physicians and Patients Alike Reconsider the Popularity of This Herbal Remedy for BPH symptoms." *Journal of Family Practice* 61(7): 418–420.

Shi, Rong, Qiungwen Xie, X. Gang et al. February 2008. "Effect of Saw Palmetto Soft Gel Capsule on Lower Urinary Tract Symptoms Associated with Benign Prostatic Hyperplasia: A Randomized Trial in Shanghai, China." *The Journal of Urology* 179(2): 610–615.

Website

MedlinePlus. www.nlm.nih.gov/medlineplus.

Sea Buckthorn Oil

Grown throughout the world, sea buckthorn, which may also be written seabuckthorn, is a shrub that contains a wide variety of nutrients. Sea buckthorn oil is made from the seeds or the pulp.

Sea buckthorn oil has carotenoids, such as zeaxanthin and lycopene, and tocopherols, sterols, folate, and vitamins B_1, B_2, B_5, B_6, and C. In addition, sea buckthorn oil has large amount of essential fatty acids omega-3s, omega-6s, omega-7s, and omega-9s, and it has flavonoids such as kaempferol and quercetin. Sea buckthorn oil even has high concentrations of palmitoleic acid, which lubricates the skin. It is not surprising that sea buckthorn oil is said to be useful for many different medical problems, such as digestive disorders, respiratory problems, and skin ailments such as infections, wounds, psoriasis, lesions, eczema, and skin damage.[1]

To many, sea buckthorn oil is "the Swiss Army knife of supplements." It may be found in a wide range of products such as "body oils, creams, soaps and shampoos, juices, jams, candies, elixirs, wine and beer."[2] Though sea buckthorn products are becoming more widely available, they may be difficult to find in traditional stores. Still, they are easily purchased in health stores and online. The supplements tend to be expensive.

MAY SUPPORT CARDIOVASCULAR HEALTH

In a study published in 2007 in *Phytomedicine: International Journal of Phytotherapy & Phytopharmacology*, researchers from India investigated the use of sea buckthorn oil to support cardiovascular health in rabbits. They began by dividing 20 male healthy white rabbits into four groups of five rabbits. The rabbits in the first group served as the controls. The rabbits in the second group received a 1mL daily dose of sea buckthorn oil for 18 days. The third group ate a high-cholesterol diet for 60 days; the rabbits in the fourth group ate a high-cholesterol diet for 30 days and then a high-cholesterol diet and 1mL daily doses of sea buckthorn oil for an additional 30 days. The researchers found that the feeding of the high-cholesterol diets to the rabbits in groups 3 and 4 resulted in increases in total cholesterol, triglycerides, and low-density lipoprotein (LDL or "bad" cholesterol). After 30 days of a high-cholesterol diet, their levels of high-density lipoprotein (HDL or "good" cholesterol) declined. Moreover, "administration of SBT [sea buckthorn] seed oil in 30 days after cholesterol feeding restricted further rise of TC [total cholesterol] on days 45 and 60 of treatment and caused a significant decline in TG [triglyceride] levels on day 60 of observations." Overall, the researchers learned that the consumption of sea buckthorn oil by normal rabbits caused a reduction in LDL cholesterol of about 24% and an increase in HDL cholesterol of about 13%. They concluded that their findings indicate that sea buckthorn oil "has significant anti-atherogenic activity [activity that helps prevent the formation of plaque] when administered to normal or hypercholesterolemic animals."[3]

ANTIBACTERIAL AND ANTIFUNGAL PROPERTIES

In a study published in 2011 in the *Journal of Plant Pathology & Microbiology*, researchers from India conducted in vitro testing to determine the antibacterial and antifungal properties of sea buckhorn oil, leaves, and seed extract. And, their results were compared to similar tests conducted with the antibacterial medication Kanamycin and the antifungal medication Clotrimazole, which served as controls.

The researchers learned that sea buckthorn oil has "several strong antioxidative and antimicrobial properties, which are due to high content of tocopherols, tocotrienols, and carotenoids present in the oil." The researchers concluded that sea buckthorn oil "is a potential source of bioactive antimicrobial agents, which could be used as a natural preservative and for nutraceutical formulations."[4]

MAY BE USEFUL FOR DRY EYES

In a double-blind, randomized, parallel trial published in 2010 in *the Journal of Nutrition*, researchers from Finland wanted to determine if sea buckthorn oil would help people who suffer from dry eyes. They began with a cohort of 100 men and women between the ages of 20 and 75 who had dry eye symptoms. For three months, every day, the participants consumed either 2g of sea buckthorn oil made from both the pulp and the seeds or a placebo. The participants kept logbooks in which they described their dry eye symptoms. Eighty-six percent of the participants completed the study. The researchers found that consumption of sea buckthorn oil has the ability to "attenuate the increase in tear film osmolarity [salt content in eyes that tends to be higher in people with dry eyes] during the cold season." Furthermore, sea buckthorn oil might be useful for eye redness, and "contact lens wearers reported fewer overall eye symptom days in the SB [sea buckthorn] group."[5]

MAY BE USEFUL FOR STOMACH DISCOMFORT

In a study published in 2012 in the *Journal of Medicinal Plants Research*, researchers from China and Finland wanted to learn if sea buckthorn oil would be useful for stomach discomfort. They began by randomly dividing 36 rats into four groups of 9 rats. One group was treated with distilled water; the rats in that group served as the controls. The rats in a second group were treated with cimetidine, a medication used for gastroesophageal reflux disease, in which there is the backward flow of acid from the stomach into the esophagus or other illnesses that cause the stomach to produce too much acid. The third and fourth groups were treated with two different doses of sea buckthorn pulp oils. After several hours, the rats were sacrificed and stomachs and gastric contents were removed and examined. Tests were also administered on mice to understand the effects that sea buckthorn oil had on gastric emptying. The researchers found that sea buckthorn oil reduced gastric secretions and delayed gastric emptying. They noted that their findings indicate that sea buckthorn pulp oil may be effective for "the treatment of stomach discomfort and gastric ulcers."[6]

HELPS HEAL WOUNDS

In a study published in 2009 in *Food and Chemical Toxicology*, researchers from India investigated the "safety and efficacy" of using sea buckthorn oil on experimentally administered wounds on male Sprague-Dawley rats. The rats were divided into two groups. The rats in the first group had their wounds treated with various doses of oral and topical sea buckthorn oil; the wounds of the rats in the second group were treated with various doses of topical sea buckthorn oil. The study continued for seven days. Some rats were put aside to serve as controls. They received no treatment. The researchers found that sea buckthorn oil was a safe and effective treatment for wounds. For example, "the histological examinations showed that tissue regeneration was much better in the SBT [sea buckthorn] seed oil treated burn wounds." They concluded that sea buckthorn seed oil "has significant wound healing activity in full-thickness burn wounds and [was] found to be safe for use."[7]

In a study published in 2006 in *Nan Fang Yi Da Xue Xue Bao* (*Journal of Southern Medical University*), researchers from China tested the use of sea buckthorn oil for healing wounds of 151 patients with burns. They used the sea buckthorn oil as an "inner dressing" and covered it with a "disinfecting dressing." The controls had their wound treated with a Vaseline gauze. The findings were notable. "The patients receiving the dressing showed more obvious exudation [oozing] reduction, pain relief, and faster epithelial cell growth and wound healing, with statistically significant difference between the two groups." The researchers concluded that sea buckthorn oil "has definite effects on the healing of burn wounds."[8]

MAY BE USEFUL FOR OVERWEIGHT AND OBESE WOMEN

In a randomized, crossover study published in 2011 in the *European Journal of Clinical Nutrition*, researchers from Finland included different preparations of sea buckthorn and bilberries in the diets of 110 female overweight and obese women. They wanted to learn if these products would reduce their increased risk for metabolic disorders, such as type 2 diabetes. For four different periods of 33–35 days, the women consumed a diet containing sea buckthorn oil or sea buckthorn extract or sea buckthorn berry or bilberries. Each study period was followed by a washout period of 30–39 days. Eighty women completed the study. The researchers found that the "different berries and berry fractions have various but slightly positive effects on the associated various of metabolic diseases."[9]

MAY HELP PROTECT THE LIVER

In a study published in 2009 in *Food and Chemical Toxicology*, researchers from Taiwan attempted to learn if sea buckthorn had liver-protective properties in mice. The researchers began by using carbon tetrachloride to induce liver damage in the mice. For eight weeks, they treated the various groups of mice with different amounts of sea buckthorn seed oil. The control mice were given olive oil. The researchers found that sea buckthorn oil had protected the liver from damage. And, they concluded that sea buckthorn oil "may be useful as a hepatoprotective agent against chemical-induced hepatotoxicity in vivo."[10]

IS SEA BUCKTHORN OIL BENEFICIAL?

Though there is not a lot of research on the benefits of sea buckthorn oil, large numbers of people maintain that it has been useful for a wide variety of medical problems. Hopefully, future research will yield more definitive results.

NOTES

1. *Better Nutrition.* www.betternutrition.com.

2. Erickson, Kim. January 2010. "Secrets of Sea Buckthorn: Learn Why This Natural Supplement Is Great for Skin, Heart Health, and More." *Better Nutrition* 72(1): 26.

3. Basu, M.R. Prasad, P. Jayamurthy et al. November 2007. "Anti-Atherogenic Effects of Seabuckthorn (*Hippophae rhamnoides*) Seed Oil." *Phytomedicine: International Journal of Phytotherapy and Phytopharmacology* 14(11): 770–777.

4. Gupta, Sanjay Mohan, Atul K. Gupta, Zakwan Ahmed, and Anil Kumar. 2011. "Antibacterial and Antifungal Activity in Leaf, Seed Extract and Seed Oil of Seabuckthorn (*Hippophae salicifolia* D. Don)." *Journal of Plant Pathology & Microbiology* 2(2): 105.

5. Larmo, P.S., R.L. Järvinen, N.L. Setälä et al. August 2010. "Oral Sea Buckthorn Oil Attenuates Tear Film Osmolarity and Symptoms in Individuals with Dry Eyes." *The Journal of Nutrition* 140(8): 1462–1468.

6. Xing, Jianfeng, Sun Jinyao, Sasa Hu et al. April 30, 2012. "Effects of Sea Buckthorn (*Hippophae rhamnoides* L.) Pulp Oils on the Gastric Secretion, Gastric Emptying and Its Analgesic Activity." *Journal of Medicinal Plants Research* 6(16): 3240–3245.

7. Upadhyay, N.K., R. Kumar, S.K. Mandotra et al. 2009. "Safety and Healing Efficacy of Sea Buckthorn (*Hippophae rhamnoides* L.) Seed Oil on Burn Wounds in Rats." *Food and Chemical Toxicology* 47(6): 1146–1153.

8. Wang, Z.Y., X.L. Luo, and C.P. He. January 2006. "Management of Burn Wounds with *Hippophae rhamnoides* Oil." *Nan Fang Yi Ke Xue Xue Bao (Journal of Southern Medical University)* 26(1): 124, 125.

9. Lehtonen, H.M., J.P. Suomela, R. Tahvonen et al. March 2011. "Different Berries and Berry Fractions Have Various but Slightly Positive Effects on the Associated Variables of Metabolic Diseases on Overweight and Obese Women." *European Journal of Clinical Nutrition* 65(3): 394–401.

10. Hsu, Yu-Wen, Chia-Fang Tsai, Wen-Kang Chen, and Fung-Jou Lu. 2009. "Protective Effects of Seabuckthorn (*Hippophae rhamnoides* L.) Seed Oil against Carbon Tetrachloride-Induced Hepatotoxicity in Mice." *Food and Chemical Toxicology* 47(9): 2281–2288.

REFERENCES AND RESOURCES

Magazines, Journals, and Newspapers

Basu, M., R. Prasad, P. Jayamurthy et al. November 2007. "Anti-Atherogenic Effects of Seabucthorn (*Hippophae rhamnoides*) Seed Oil." *Phytomedicine: International Journal of Phytotherapy and Phytopharmacology* 14(11): 770–777.

Erickson, Kim. January 2010. "Secrets of Sea Buckthorn: Learn Why This Natural Supplement Is Great for Skin, Heart Health, and More." *Better Nutrition* 72(1): 26.

Gupta, Sanjay Mohan, Atul K. Gupta, Zakwan Ahmed, and Anil Kumar. 2011. "Antibacterial and Antifungal Activity in Leaf, Seed Extract and Seed Oil of Seabuckthorn (*Hippophae salicifolia* D. Don)." *Journal of Plant Pathology & Microbiology* 2(2): 105.

Hsu, Yu-Wen, Chia-Fang Tsai, Wen-Kang Chen, and Fung-Jou Lu. 2009. "Protective Effects of Seabuckthorn (*Hippophae rhamnoides* L.) Seed Oil against Carbon Tetrachloride-Induced Hepatotoxicity in Mice." *Food and Chemical Toxicology* 47(9): 2281–2288.

Larmo, P.S., R.L. Järvinen, N.L. Setälä et al. August 2010. "Oral Sea Buckthorn Oil Attenuates Tear Film Osmolarity and Symptoms in Individuals with Dry Eye." *The Journal of Nutrition* 140(8): 1462–1468.

Lehtonen, H.M., J.P. Suomela, R. Tahvonen et al. March 2011. "Different Berries and Berry Fractions Have Various but Slightly Positive Effects on the Associated Variables of Metabolic Diseases on Overweight and Obese Women." *European Journal of Clinical Nutrition* 65(3): 394–401.

Upadhyay, N.K., R. Kumar, S.K. Mandotra et al. 2009. "Safety and Healing Efficacy of Sea Buckthorn (*Hippophae rhamnoides* L.) Seed Oil on Burn Wounds in Rats." *Food and Chemical Toxicology* 47(6): 1146–1153.

Wang, Z.Y., X.L. Luo, and C.P. He. January 2006. "Management of Burn Wounds with *Hippophae rhamnoides* Oil." *Nan Fang Yi Ke Da Xue Xue Bao (Journal of Southern Medical University)* 26(1): 124, 125.

Xing, Jianfeng, Sun Jinyao, Sasa Hu et al. April 30, 2012. "Effects of Sea Buckthorn (*Hippophae rhamnoides*) Pulp Oils on the Gastric Secretion, Gastric Emptying and Its Analgesic Activity." *Journal of Medicinal Plants Research* 6(16): 3240–3245.

Websites

Better Nutrition. www.betternutrition.com.
International Seabuckthorn Association. www.isahome.net.

Sesame Oil

Derived from sesame seeds, sesame oil has long been considered to have powerful therapeutic properties. Widely used in India, the Middle East, East Asia, and Africa, it contains high amounts of phytosterols, such as beta-sitosterol, and phytochemical lignans, such as sesamolin and sesamin. It is believed to support cardiovascular health and help regulate sugar. It is also thought to have antioxidant and antiaging properties. Some contend that it is useful for killing bacteria and fungi.[1]

Sesame seed oil is sold in a lighter version that has a high smoke point and is suitable for deep-frying. It is also sold in a darker version that has a lower smoke point and is more suitable for stir-frying. It is easy to find in health food stores and online; it may not be available in traditional supermarkets. It is inexpensive.

ENHANCING CARDIOVASCULAR HEALTH

In a study published in 2007 in the *Journal of Agricultural and Food Chemistry*, researchers from India examined the ability of a diet containing coconut oil or coconut oil mixed with sesame oil or coconut oil mixed with rice bran oil to lower lipid levels in rats. The researchers began by dividing male Wistar rats into various groups. For 60 days, the rats were fed diets that contained 10% fat from coconut oil, sesame oil, rice bran oil, or coconut oil combined with sesame oil or rice bran oil. When compared to the rats fed just coconut oil, the rats fed blended coconut oil and sesame oil had serious reductions in serum total cholesterol, low-density lipoprotein cholesterol (LDL or "bad" cholesterol), and triacylglycerols. The researchers noted that their study "suggests that feeding fats containing [the] blended oils with balanced fatty acids lowers serum and liver lipids."[2]

In a two-phase study published in 2013 in the *European Journal of Preventive Cardiology*, researchers from Greece and Texas wanted to assess the short- and long-term effect that sesame oil had on endothelial functioning in men with hypertension (elevated blood pressure). In the first phase of the study, 26 men consumed either 35g of sesame oil or 35g of a control oil. In the second phase of the study, 30 men consumed 35g of sesame oil or 35g of a control oil every day for two months. The researchers found that the endothelial functioning improved after both the short- and long-term use of sesame oil. They commented that "this is the first study to show that sesame oil consumption exerts a beneficial effect on endothelial function and this effect is sustained with long-term daily use."[3]

In a study published in 2010 in *Redox Report*, researchers from India investigated the ability of sesamol (a compound in sesame oil) to protect the hearts of adult male albino rats that had experimentally induced heart attacks. The researchers began by dividing their rats into six groups. The rats in one group served as the controls; the rats in a second group were fed sesamol. Rats in the four remaining groups all had heart attacks induced by isoproterenol; the rats in three of these groups were fed different amounts of sesamol for nine days. While analyzing their results, the researchers found that "sesamol at a dose of 50mg/kg body weight had more cardioprotective effect than the other two doses (100mg and 200mg)."[4]

Meanwhile, in a study published in 2010 in *Lipids in Health and Disease*, different researchers from Greece compared the ability of sesame oil and *N*-acetylcysteine (NAC), an antioxidant compound naturally found in such foods as garlic and onions, to lower elevated cholesterol levels in mice. The researchers divided 25 mice into four experimental groups. One group of five mice served as the controls. These mice ate regular mice food. A second group of five mice ate a normal diet supplemented with cholesterol. The third group of seven mice ate a high-cholesterol diet supplemented with NAC. And, the fourth group of eight mice ate a high-cholesterol diet supplemented with 10% sesame oil. The trial continued for eight weeks. While the NAC proved to have cholesterol-lowering properties, the sesame oil did not display similar characteristics. According to the researchers, "it did not show significantly hypolipidemic activity."[5]

MAY BE USEFUL FOR PEOPLE WITH TYPE 2 DIABETES

In a study published in 2011 in *Clinical Nutrition*, researchers from India and Oman evaluated the use of sesame oil and the antidiabetic medication gliben- clamide for treating people with mild to moderate type 2 diabetes. The research- ers began by dividing the cohort, which consisted of 32 men and 28 women with type 2 diabetes, into three groups. The 18 members of the first group received sesame oil, the 20 members of the second group received glibenclamide, and the 22 members of the third group received both the sesame oil and the medication. At the end of the 60-day trial, the researchers found that the best response was obtained by the combined sesame oil/medication therapy. The researchers noted that their study "demonstrated that glibenclamide in combination with sesame oil has significantly enhanced anti-hyperglycemic efficacy as compared with their monotherapy." According to these researchers, sesame oil by itself "only had moderate lipid lowering effect."[6]

WOUND-HEALING PROPERTIES

In a study published in 2008 in the *Indian Journal of Experimental Biology*, re- searchers from India tested the wound-healing properties of sesame seeds and sesame oil in male albino rats. The researchers began by creating several different types of wounds on rats and then treating them topically with aloe vera or two different doses of sesame seeds in a gel and sesame oil—2.5% or 5%. The rats were also treated orally with aloe vera, sesame seeds, and sesame oil. The researchers wrote that both the sesame seeds in a gel and the sesame oil applied topically or administered orally "promoted the breaking strength, wound contraction and pe- riod of epithelization in different models of experimental wounds." That means that both the seeds and oil enhanced the healing process. In addition, the re- searchers added that "the low dose of both seeds and oil are more effective when applied locally and the high dose of seeds and oil showed greater effect in dead space wound when administered orally."[7]

In a study published in 2011 in the *Journal of Ethnopharmacology*, researchers from India investigated the use of sesamol to treat wounds in normal albino rats and in rats treated with dexamethasone, a corticosteroid used to impair healing. The researchers began by dividing male and female albino rats into four groups.

Under anesthesia, wounds were inflicted on all of the rats. The rats in the first group were the controls, the rats in the second group were treated with sesamol, the rats in the third group were treated with dexamethasone, and the rats in the final group were treated with sesamol and dexamethasone. The researchers determined that sesamol "is a promising entity which promotes the wound healing in both normal and delayed healing conditions."[8]

SUPPORTS HEALTH

In a study published in 2011 in *Drug & Chemical Toxicology*, researchers from India wanted to determine the protective effects of sesame oil on oxidative DNA damage and lipid peroxidation caused by 4-Nitroquinoline-1-oxide (4-NQO), a substance used to cause tumors in laboratory animals. The researchers began by dividing 42 male and female Wistar rats into seven groups. The rats in the first group served as the controls, the rats in the second group ate sesame oil, and the rats in the third group were injected with 4-NQO. The rats in the final four groups were all injected with 4-NQO and were given different doses of sesame oil. After 24 hours, the researchers determined that "pretreatment with sesame oil effectively protected against DNA damage in a dose-dependent fashion." In fact, the highest oxidative DNA protection was observed at the highest dose of sesame oil. And, they concluded that "the antioxidant sesame oil effectively protected DNA damage and LPO [lipid peroxidation] induced by 4-NQO."[9]

IS SESAME OIL BENEFICIAL?

Sesame oil appears to have some benefits. Many more people may want to add it to their everyday oils.

NOTES

1. Olivier, Rachel. December 2011. "An Overview of the Beneficial Attributes of Sesame Oil." *Original Internist* 18(4): 133–139.

2. Reena, Malongil B. and Belur R. Lokesh. 2007. "Hypolipidemic Effects of Oils with Balanced Amounts of Fatty Acids Obtained by Blending and Interesterification of Coconut Oil and Rice Bran oil or Sesame Oil." *Journal of Agricultural and Food Chemistry* 55(25): 10,461–10,469.

3. Karatzi, K., K. Stamatelopoulos, M. Lykka et al. 2013. "Sesame Oil Consumption Exerts a Beneficial Effect on Endothelial Function in Hypertensive Men." *European Journal of Preventive Cardiology* 20(2): 202–208.

4. Vennila, L. and K. V. Pugalendi. 2010. "Protective Effect of Sesamol against Myocardial Infarction Caused by Isoproterenol in Wistar Rats." *Redox Report* 15(1): 36–42.

5. Agrogiannis, George, Dimitrios Iliopoulos, Theodoros Karatzas et al. 2010. "Comparative Antilipidemic Effect of N-Acetylcysteine and Sesame Oil Administration in Diet-Induced Hypercholesterolemic Mice." *Lipids in Health and Disease* 9(1): 23–29.

6. Sankar, Devarajan, Amanat Ali, Ganapathy Sambandam, and Ramakrishna Rao. 2011. "Sesame Oil Exhibits Synergistic Effect with Anti-Diabetic Medication in Patients with Type 2 Diabetes Mellitus." *Clinical Nutrition* 30(3): 351–358.

7. Kiran, Kotade and Mohammed Asad. November 2008. "Wound Healing Activity of *Sesamum indicum* L. Seed and Oil in Rats." *Indian Journal of Experimental Biology* 46(): 777–782.

8. Shenoy, Rekha R., Arun T. Sudheendra, Pawan G. Nayak et al. 2011. "Normal and Delayed Wound Healing Is Improved by Sesamol, an Active Constituent of *Sesamum indicum* (L.) in Albino Rats." *Journal of Ethnopharmacology* 133(2): 608–612.

9. Arumugam, Ponnan and Samiraj Ramesh. April 2011. "Protective Effects of Sesame Oil on 4-NQO-Induced Oxidative DNA Damage and Lipid Peroxidation in Rats." *Drug & Chemical Toxicology* 34(2): 116–119.

REFERENCES AND RESOURCES

Magazines, Journals, and Newspapers

Agrogiannis, George, Dimitrios Iliopoulos, Theodoros Karatzas et al. 2010. "Comparative Antilipidemic Effect of N-Acetylcysteine and Sesame Oil Administration in Diet-Induced Hypercholesterolemic Mice." *Lipids in Health and Disease* 9(1): 23–29.

Arumugam, Ponnan and Samiraj Ramesh. April 2011. "Protective Effects of Sesame Oil on 4-NQO-Induced Oxidative DNA Damage and Lipid Peroxidation in Rats." *Drug & Chemical Toxicology* 34(2): 116–119.

Karatzi, K., K. Stamatelopoulos, M. Lykka et al. April 2013. "Sesame Oil Consumption Exerts a Beneficial Effect on Endothelial Function in Hypertensive Men." *European Journal of Preventive Cardiology* 20(2): 202–208.

Kiran, Kotade and Mohammed Asad. November 2008. "Wound Healing Activity of *Sesamum indicum* L. Seed and Oil in Rats." *Indian Journal of Experimental Biology* 46: 777–782.

Olivier, Rachel. December 2011. "An Overview of the Beneficial Attributes of Sesame Oil." *Original Internist* 18(4): 133–139.

Reena, Malongil and Belur R. Lokesh. 2007. "Hypolipidemic Effect of Oils with Balanced Amounts of Fatty Acids Obtained by Blending and Interesterification of Coconut Oil with Rice Bran Oil or Sesame Oil." *Journal of Agricultural and Food Chemistry* 55(25): 10,461–10,469.

Sankar, Devarajan, Amanat Ali, Ganapathy Sambandam, and Ramakrishna Rao. 2011. "Sesame Oil Exhibits Synergisic Effect with Anti-Diabetic Medication in Patients with Type 2 Diabetes Mellitus." *Clinical Nutrition* 30(3): 351–358.

Shenoy, Rekha R., Arun T. Sudheendra, Pawan G. Nayak et al. 2011. "Normal and Delayed Wound Healing Is Improved by Sesamol, an Active Constituent of *Sesamum indicum* (L.) in Albino Rats." *Journal of Ethnopharmacology* 133(2): 608–612.

Vennila, L. and K. V. Pugalendi. 2010. "Protective Effect of Sesamol against Myocardial Infarction Caused by Isoproterenol in Wistar Rats." *Redox Report* 15(1): 36–42.

Website

LIVESTRONG.COM. www.livestrong.com.

Sunflower Oil

Made from sunflower seeds, there are three main types of sunflower oil. High-oleic sunflower oil contains at least 79% monounsaturated fat. Its composition is very similar to olive oil. Mid-oleic sunflower oil has 55%–75% monounsaturated fats. This type is frequently used for commercial cooking. And, linoleic sunflower oil has about 65% polyunsaturated fat. It is usually used in salad dressings and margarine. (Oleic acid is a monounsaturated fatty acid found naturally in many plant sources and animal products.) Both monounsaturated and polyunsaturated fats are believed to be healthful. They are thought to reduce inflammation, lower low-density lipoprotein (LDL or "bad" cholesterol), and raise high-density lipoprotein (HDL or "good" cholesterol).[1] Sunflower oil is light in appearance and taste, and it has very high amounts of vitamin E. Somewhat expensive, it is readily available in supermarkets and online.

SUPPORTS CARDIOVASCULAR HEALTH

In a study published in 2011 in the *British Journal of Nutrition*, researchers from the United Kingdom noted that insulin-resistant-related conditions, such as obesity and type 2 diabetes, are characterized by metabolic abnormalities, including elevated levels of nonesterified fatty acids (NEFA) in the blood. They designed a study that tested the effect of consuming drinks containing rich amounts of saturated fatty acids from palm stearin or rich amounts of monounsaturated fatty

Vitamin D tablets with a sunflower.
(Boomfeed/Dreamstime.com)

acids from high-oleic sunflower oil on levels of serum NEFA. There was also a control drink that had no fat. "The composition of fat consumed is an important factor in determining the impact of elevated NEFA on both vascular function and endothelial inflammation."

The cohort consisted of 10 healthy men. They consumed the drinks on three different occasions—separated by at least one week. The researchers found that the consumption of the saturated fat drink resulted in the acute elevation of serum NEFA, which lead to cardiovascular problems such as arterial stiffness. "These effects were not found following ingestion of a MUFA [monounsaturated fatty acids]-rich drink." The researchers commented that their "protocol resulted in a three to four increase in NEFA from baseline in our healthy subjects, achieving levels observed in insulin resistant states such as obesity and T2D [type 2 diabetes]."[2]

In a study published in 2010 in the *Journal of Agricultural and Food Chemistry*, researchers from Italy wanted to determine if micronutrient-enriched high-oleic sunflower oils could have a beneficial effect on the plasma lipid profile and antioxidant status of Sprague-Dawley rats fed a high-fat diet. For four weeks, they divided rats into several groups and fed them the same basic foods supplemented with sunflower oils that contained varying types and amounts of micronutrients. (This is a result of "the different crushing and refining procedures used.") The rats in the control group received refined high-oleic sunflower oil. The researchers found that the optimized oils supported cardiovascular health. "The reduction in plasma triglyceride and total cholesterol levels was 43% and 20%, respectively, in the group fed the diet with the highest levels of micronutrients."

The researchers noted that their study "demonstrates that improved oil refining process produce optimized oils that are able to exert beneficial effects on CVD [cardiovascular disease] factors, enhancing the hypolipidemic effects of high-oleic acid sunflower oil and the antioxidant defense mechanisms."[3]

In a randomized, double-blind, crossover study published in 2011 in *The Journal of Nutrition*, researchers from the Netherlands compared the "acute effects" of a breakfast rich in saturated fatty acids from butter versus a breakfast rich in polyunsaturated fatty acids from sunflower oil. Though the sources of fat differed, both meals contained 50g fat. The initial cohort consisted of 15 overweight men; of these, 14 completed the study. All of the participants ate both meals, on two different occasions, separated by at least seven days. The researchers found that substituting sunflower oil for butter "may decrease postprandial lipemia [the elevation of lipid levels following a meal] markers of inflammation and endothelial activation in overweight men."[4]

In a study published in 2011 in the *Saudi Medical Journal*, researchers from India and Saudi Arabia examined the effect of the consumption of a combination of sunflower oil and sesame oil on people with hypertension (elevated blood pressure). The researchers began by separating men, between the ages of 45 and 55, into groups. The first group consisted of 14 men with hypertension; the second group consisted of 38 men on the calcium channel blocker medication nifedipine. The second group was then divided into two groups. Twelve men continued to take nifedipine and 26 men took nifedipine plus a combination of sunflower and sesame oils. The men on the combination of nifedipine and oils were told to use only those oils for the duration of the study. After 45 days, the researchers found that men on the medication/oil combination mixture had "better protection [from high blood pressure] compared to nifedipine alone." Moreover, the researchers added, "the treatment with oil-mix further possessed anti hyperlipidemic and antioxidant properties besides normalizing the electrolytes in patients with hypertension." They concluded that "the combination of sesame and sunflower oil mix reduces the risk of cardiovascular disease."[5]

A little less laudatory results were obtained from a study published in 2009 in *Nutrición Hospitalaria*. Researchers from Madrid, Spain, tested the "modulating effect" of sunflower oil and olive oil on the hypolipidemic effect of the statin medication simvastatin. The cohort consisted of 25 men, between the ages of 45 and 65, with "severe hypercholesterolemia." For six months, 13 of the men used sunflower oil as their culinary oil and 12 used olive oil. While the participants in both groups experienced reductions in their total cholesterol and LDL cholesterol, the overall results were better for olive oil group than the sunflower oil group. For example, the participants in the olive oil group had a greater decrease in the total cholesterol/HDL ratio than the participants in the sunflower oil group. The researchers noted "olive-oil diets in preference to sunflower-oil diets must be consumed in hypercholesterolemic subjects with Simvastatin."

Another study on sunflower oil and olive oil was published in 2012 in *BMJ*. In the study, which was conducted in Spain, the researchers examined any association between the consumption of food fried in sunflower and olive oils and cardiovascular disease. The cohort consisted of 40,757 adults between the ages of 29 and 69 who were free of heart disease at baseline. During the 11-year follow-up period, there were 606 events linked to heart disease and 1,134 deaths. The researchers learned that, on average, 138g of fried food was consumed daily. This included 14g of oil used for frying. About 7% of the total amount consumed was fried. It is important to note that the results "did not vary between those who used olive oil for frying and those who used sunflower oil." And, no association was observed between the consumption of fried foods and coronary heart disease and other causes of mortality.[6]

MAY HELP MAINTAIN WEIGHT

In a randomized, crossover study published in 2008 in *Metabolism Clinical and Experimental*, researchers from Canada wondered if the type of fat people consumed—olive oil, sunflower oil, or flaxseed oil—could influence their long-term ability to manage their weight. The cohort consisted of 15 healthy male university students. Separated by a week of regular eating, the students ate breakfasts that contained large amounts of each type of oil. Except for the type of oil, the meals were identical. The researchers found that olive oil used the most energy.

But, sunflower oil was second. So, it is possible that including sunflower oil in the diet may help people with weight management.[7]

IS SUNFLOWER OIL BENEFICIAL?

Sunflower oil seems to have a number of benefits. And, since it is relatively inexpensive, it is easy to incorporate into the diet. However, other oils appear to have more overall health benefits. So, sunflower oil should be used in conjunction with other oils.

NOTES

1. LIVESTRONG.COM. www.livestrong.com.

2. Newens, K.J., A.K. Thompson, K.G. Jackson et al. May 2011. "Acute Effects of Elevated NEFA on Vascular Function: A Comparison of SFA and MUFA." *British Journal of Nutrition* 105(9): 1343–1351.

3. Di Benedetto, R., L. Attorri, F. Chiarotti et al. May 12, 2010. "Effect of Micronutrient-Enriched Sunflower Oils on Plasma Lipid Profile and Antioxidant Status in High-Fat-Fed Rats." *Journal of Agricultural and Food Chemistry* 58(9): 5328–5333.

4. Masson, C.J. and R.P. Mensink. May 2011. "Exchanging Saturated Fatty Acids for (n-6) Polyunsaturated Fatty Acids in a Mixed Meal May Decrease Postprandial Lipemia and Markers of Inflammation and Endothelial Activity in Overweight Men." *The Journal of Nutrition* 141(5): 816–821.

5. Sudhakar, B., P. Kalaiarasi, K.S. Al-Numair et al. April 2011. "Effect of Combination of Edible Oils on Blood Pressure, Lipid Profile, Lipid Peroxidative Markers, Antioxidant Status, and Electrolytes in Patients with Hypertension on Nifedipine Treatment." *Saudi Medical Journal* 32(4): 379–385.

6. Guallar-Castillón, Pilar, Fernando Rodriguez-Artalejo, Esther Lopez-Garcia et al. 2012. "Consumption of Fried Foods and Risk of Coronary Heart Disease: Spanish Cohort of the European Prospective Investigation into Cancer and Nutrition Study." *BMJ* 344: e363.

7. Jones, Peter J.H., Stephanie Jew, and Suhad AbuMweis. September 2008. "The Effect of Dietary Oleic, Linoleic, and Linolenic Acids on Fat Oxidation and Energy Expenditure in Healthy Men." *Metabolism Clinical and Experimental* 57(9): 1198–1203.

REFERENCES AND RESOURCES

Magazines, Journals, and Newspapers

Bester, D., A.J. Esterhuyse, E.J. Truter, and J. van Rooyen. December 2010. "Cardiovascular Effects of Edible Oils: A Comparison between Four Popular Edible Oils." *Nutrition Research Reviews* 23(2): 334–348.

Di Benedetto, R., L. Attorri, F. Chiarotti et al. May 12, 2010. "Effect of Micronutrient-Enriched Sunflower Oils on Plasma Lipid Profile and Antioxidant Status in High-Fat-Fed Rats." *Journal of Agricultural and Food Chemistry* 58(9): 5328–5333.

Guallar-Castillón, Pilar, Fernando Rodriguez-Artalejo, Esther Lopez-Garcia et al. 2012. "Consumption of Fried Foods and Risk of Coronary Heart Disease: Spanish Cohort of the European Prospective Investigation into Cancer and Nutrition Study." *BMJ* 344: e363.

Jones, Peter J. H., Stephanie Jew, and Suhad AbuMweis. September 2008. "The Effect of Dietary Oleic, Linoleic, and Linolenic Acids on Fat Oxidation and Energy Expenditure in Healthy Men." *Metabolism Clinical and Experimental* 57(9): 1198–1203.

Masson, C. J. and R. P. Mensink. May 2011. "Exchanging Saturated Fatty Acids for (n-6) Polyunsaturated Fatty Acids in a Mixed Meal May Decrease Postprandial Lipemia and Markers of Inflammation and Endothelial Activity in Overweight Men." *The Journal of Nutrition* 141(5): 816–821.

Newens, K. J., A. K. Thompson, K. G. Jackson et al. May 2011. "Acute Effects of Elevated NEFA on Vascular Function: A Comparison of SFA and MUFA." *British Journal of Nutrition* 105(9): 1343–1351.

Sánchez-Muniz, F. J., S. Bastida, O. Gutiérrez-García, and A. Carbajal. May–June 2009. "Olive Oil-Diet Improves the Simvastatin Effects with Respect to Sunflower Oil-Diet in Men with Increased Cardiovascular Risk: A Preliminary Study." *Nutrición Hospitalaria* 24(3): 333–339.

Sudhakar, B., P. Kalaiarasi, K. S. Al-Numair et al. April 2011. "Effect of Combination of Edible Oils on Blood Pressure, Lipid Profile, Lipid Peroxidative Markers, Antioxidant Status, and Electrolytes in Patients with Hypertension on Nifedipine Treatment." *Saudi Medical Journal* 32(4): 379–385.

Websites

LIVESTRONG.COM. www.livestrong.com.
National Sunflower Association. www.sunflowernsa.com.

Walnut Oil

Walnuts may be traced to 7000 BC. In fact, they are the oldest known tree food. Today, walnuts are widely available, and they are sometimes dried and cold-pressed into oil. Topaz in color, walnut oil has a delicious nutty taste. It is a rich source of antioxidants, such as ellagic acid as well as manganese, copper, fiber, and alpha-linolenic acid, an omega-3 fatty acid. Walnut oil also has good amounts of melatonin. It is thought to support cardiovascular health, lower inflammation, and reduce the risk for cancer. Furthermore, walnut oil may help to correct skin problems such as eczema.[1] Walnut oil is readily available in specialty stores and online. Since it is expensive, it may be harder to find in more traditional supermarkets.

CARDIOVASCULAR HEALTH

In a randomized, crossover study published in 2010 in the *Journal of the American College of Nutrition*, researchers from Pennsylvania (United States) and Ontario (Canada) wanted to learn how diets that included walnut oil, walnuts, and flaxseed oil could improve the cardiovascular health of people with elevated levels of serum cholesterol. They began by dividing their cohort, which consisted of 20 subjects, into three groups. All of the subjects consumed one of three six-week diets in a random order. The first diet was a typical "American" diet that included no nuts, the second diet included 1.3 ounces of walnuts and a tablespoon of walnut oil, and the third diet included walnuts, walnut oil, and 1.5 tablespoons of flaxseed oil. Following each diet, the subjects had two stressful tests. In the first test, they had only two minutes to prepare a three-minute speech; in the second, they had one foot submerged in ice-cold water. During these tests, blood pressure readings were taken. The researchers found that the average diastolic blood pressure (bottom number) of the dieters who were consuming walnuts and walnut oil was significantly reduced. A subset of the subjects had vascular ultrasounds that measured artery dilation. The researchers found that adding flaxseed oil to the walnut and walnut oil diet improved this test of vascular health. They noted that their findings "suggest novel mechanisms for the cardio protective effects of walnuts and flax."[2]

In a randomized, double-blind, case-controlled study published in 2003 in *Angiology*, researchers based in Shiraz, Iran, investigated the role that walnut oil may play in lowering elevated lipid levels in the blood, a condition known as hyperlipidemia. The cohort consisted of 60 people with hyperlipidemia. The subjects were randomly placed into one of two groups. Group A, which consisted of 29 people, received encapsulated walnut oil supplementation; the 31 members of Group B took placebos. The lipid levels of the members of both groups were taken on several occasions. The researchers found that walnut oil significantly reduced lipid levels. "It was concluded that walnut oil is a good antihypertriglyceridemic natural remedy."[3]

On the other hand, in a study published in 2013 in *Thrombosis Research*, researchers from the Netherlands and Turkey tested the ability of whole walnuts and walnut oil to lower the risk of cardiovascular disease in mice fed a high-fat diet. For eight weeks, in addition to their high-fat diet, the mice received whole walnut or walnut oil or sunflower oil (control) supplementation. Their blood and livers were then analyzed for lipid content. Interestingly, the researchers found that the intake of whole walnuts but not walnut oil caused reductions in the formation of plaque and improvements in plasma lipids. "Feeding mice with walnut oil did not provoke significant changes in these parameters in comparison to the control diet."[4]

ANTIDIABETIC EFFECTS

In a study published in 2011 in the *African Journal of Pharmacy and Pharmacology*, researchers from Iran tested the ability of walnut oil to help those dealing with type 1 diabetes mellitus. They began by using alloxan to induce diabetes in 20 adult male rats. (Alloxan is an oxidized product of uric acid that destroys islet cells of the pancreas, thus causing diabetes.) The diabetic rats were then divided into two groups. One group ate an ordinary diet; the second group was treated with walnut oil. The researchers also had other nondiabetic rats that received neither alloxan nor walnut oil. At the end of the trial, which continued for six weeks, the researchers found "that walnut oil was able to reduce blood glucose significantly when compared with the control group." They concluded that walnut oil was useful "in the treatment of diabetes mellitus type 1."[5]

ANTICANCER PROPERTIES

In a study published in 2010 in *Lipids in Health and Disease*, researchers from Argentina, India, and Shaker Heights, Ohio (United States) compared the ability of walnut oil and peanut oil to protect against the growth of breast cancer. The researchers began by dividing 60 mice into three groups of 20 mice. While the mice in the control group ate a normal mouse diet, the mice in second group ate a diet with 6% walnut oil and the mice in the third group ate a diet with 6% peanut oil. After three months of consuming these diets, the mice were inoculated with tumor tissue. When compared to the control mice, the researchers found that the growth of tumor cells in the mice fed walnut oil and peanut oil was less. Both the walnut oil and peanut oil groups had fewer metastases than the control group. In addition, the walnut oil and peanut oil–fed mice survived longer than the control mice. The researchers noted that "nutritional factors play a major role in cancer initiation and development."[6]

On the other hand, Elaine Hardman, PhD, a longtime walnut researcher at the Joan C. Edwards School of Medicine at Marshall University in West Virginia, does not believe that walnut oil alone is sufficient to prevent or fight breast cancer. She maintains that one needs to eat whole walnuts in order to kill cancer cells. According to Dr. Hardman, "the oil would contain the omega-3 fat and lipid soluble components but would be missing other components that would be retained in the pulp."[7]

MAY REDUCE INCIDENCE OF ALLERGIES IN CHILDREN

In a study published in 2011 in *The Journal of Physiology*, researchers from France wanted to learn if feeding a diet higher in omega-3, such as the omega-3 contained in walnut oil, to pregnant pigs would reduce the incidence of allergies in their offspring. The researchers fed the pregnant pigs either a lard-based diet or an omega-3-based diet. When the newborn piglets were 28 days, the researchers found that the gut permeability of the omega-3-fed piglets was higher than the lard-fed piglets. When the gut is more permeable, the immune system develops faster. So, there was better immune function, which, in turn, resulted in fewer allergies. Since the pig intestine is an excellent model of the human intestine, according to the researchers, these responses to the intake of omega-3 may well occur in humans.[8]

IS WALNUT OIL BENEFICIAL?

Walnut oil certainly appears to be a desirable addition to the diet. But, according to a 2009 article in the *International Journal of Food Sciences and Nutrition*, people who use walnut oil should be aware of its limitations. After studying eight "healthier" oils (walnut, almond, avocado, hazelnut, macadamia nut, grape seed, rice bran, and toasted sesame), researchers from the United Kingdom learned that walnut oil was the most unstable; it was "very prone to oxidative deterioration." They wrote that "it is worth pointing out that once the walnut oil bottle is opened, it should be consumed within a couple of months."[9] It should never be used for cooking and, once opened, it should be stored in the refrigerator.

NOTES

1. California Walnuts Website. www.walnuts.org.
2. West, Sheila G., Andrea Likos Krick, Laura Cousino Klein et al. December 2010. "Effect of Diets High in Walnuts and Flax Oil on Hemodynamic Responses to Stress and Vascular Endothelial Function." *Journal of the American College of Nutrition* 29(6): 595–603.
3. Zibaeenezhad, M.J., M. Rezaiezadeh, A. Mowla et al. July–August 2003. "Antihypertriglyceridemic Effect of Walnut Oil." *Angiology* 54(4): 411–414.
4. Nergiz-Ünal, R., M.J. Kuijpers, S.M. de Witt et al. May 2013. "Atheroprotective Effect of Dietary Walnut Intake in ApoE-Deficient Mice: Involvement of Lipids and Coagulation Factors." *Thrombosis Research* 131(5): 411–417.
5. Rahimi, Parivash, Najmeh Kabiri, Sedigheh Asgary, and Mahbubeh Setorki. December 29, 2011. "Anti-Diabetic Effects of Walnut Oil on Alloxan-Induced Diabetic Rats." *African Journal of Pharmacy and Pharmacology* 5(24): 2655–2661.
6. Comba, Andrea, Damian M. Maestri, Maria A. Berra et al. October 8, 2010. "Effect of ω-3 and ω-9 Fatty Acid Rich Oils on Lipoxygenases and Cyclooxygenases Enzymes and on the Growth of a Mammary Adenocarcinoma Model." *Lipids in Health and Disease* 9(1): 112–122.
7. About.com Breast Cancer Website. Breastcancer.about.com.
8. De Quelen, F., J. Chevalier, M. Rolli-Derkinderen et al. September 1, 2011. "n-3 Polyunsaturated Fatty Acids in the Maternal Diet Modify the Postnatal Development of Nervous Regulation of Intestinal Permeability in Piglets." *The Journal of Physiology* 589(17): 4341–4352.
9. Kochhar, S. Parkash and C. Jeya K. Henry. 2009. "Oxidative Stability and Shelf-Life Evaluation of Selected Culinary Oils." *International Journal of Food Sciences and Nutrition* 60(Supplement 7): 289–296.

REFERENCES AND RESOURCES

Magazines, Journals, and Newspapers

Comba, Andrea, Damian M. Maestri, Maria A. Berra et al. October 8, 2010. "Effect of ω-3 and ω-9 Fatty Acid Rich Oils on Lipoxygenases and Cyclooxygenases Enzymes and on the Growth of a Mammary Adenocarcinoma Model." *Lipids in Health and Disease* 9(1): 112–122.
De Quelen, F., J. Chevalier, M. Rolli-Derkinderen et al. September 1, 2011. "n-3 Polyunsaturated Fatty Acids in the Maternal Diet Modify the Postnatal Development of Nervous Regulation of Intestinal Permeability in Piglets." *The Journal of Physiology* 589(17): 4341–4352.
Kochhar, S. Parkash and C. Jeya K. Henry. 2009. "Oxidative Stability and Shelf-Life Evaluation of Selected Culinary Oils." *International Journal of Food Sciences and Nutrition* 60(Supplement 7): 289–296.

Nergiz-Ünal, R., M. J. Kuijpers, S. M. de Witt et al. May 2013. "Atheroprotective Effect of Dietary Walnut Intake in ApoE-Deficient Mice: Involvement of Lipids and Coagulation Factors." *Thrombosis Research* 131(5): 411–417.

Rahimi, Parivash, Najmeh Kabiri, Sedigheh Asgary, and Mahbubeh Setorki. December 29, 2011. "Anti-Diabetic Effects of Walnut Oil on Alloxan-Induced Diabetic Rats." *African Journal of Pharmacy and Pharmacology* 5(24): 2655–2661.

West, Sheila G., Andrea Likos Krick, Laura Cousino Klein et al. December 2010. "Effects of Diets High in Walnuts and Flax Oil on Hemodynamic Responses to Stress and Vascular Endothelial Function." *Journal of the American College of Nutrition* 29(6): 595–603.

Zibaeenezhad, M. J., M. Rezaiezadeh, A. Mowla et al. July–August 2003. "Antihypertriglyceridemic Effect of Walnut Oil." *Angiology* 54(4): 411–414.

Websites

About.com Breast Cancer. Breastcancer.about.com.
California Walnuts. www.walnuts.org.

Wheat Germ Oil

Dark amber or brown in color, wheat germ oil is derived from wheat germ, which is found in the embryo of the wheat kernel. Wheat germ oil has a nutty scent and is thought to contain a wide variety of vitamins and minerals. These include vitamins A, D, and E and many of the B vitamins. Wheat germ oil containsiron, phosphorus, fatty acids—especially linoleic and alpha-linoleic acids—which are essential fatty acids (the body needs these to function but is unable to produce them). Wheat germ oil also has octacosanol, a chemical found in many plants that has several health benefits, such as increasing energy, endurance, and strength, reducing stress, improving heart beats, controlling muscle spasms, and reducing the pain associated witharthritis.[1] A 2009 article in the *Journal of Oleo Science* noted that wheat germ oil has natural antioxidant components "that may serve as dietary sources for natural antioxidants scavenging free radicals and thus prevent the body from diseases and promote human health." The article also maintained that wheat germ oil "has been found to reduce plasma and liver cholesterol in animals and to delay aging."[2] Another article, published in 2012 in the *International Journal of Food Sciences and Nutrition* noted that the benefits of "wheat germ and its derivatives include lowering plasma and liver cholesterol, reducing cholesterol absorption, inhibiting platelet aggregation, improving physical endurance, retarding aging, improving fertility . . ., as well as preventing and curing carcinogenesis."[3]

Wheat germ oil may be a little difficult to find in traditional stores. But, it is readily available online, and it should be sold in some specialty stores. Generally, wheat germ oil is moderately priced.

SUPPORTS CARDIOVASCULAR HEALTH

In a letter to the editor published in 2006 in *Arteriosclerosis, Thrombosis, and Vascular Biology*, researchers from the University of Rome reported on their studies of the effects of alpha-linoleic acid on people with mildly elevated levels of cholesterol. The researchers randomly divided a group of subjects, all of whom had elevated levels of cholesterol, into one of two groups. For two months, the subjects took daily supplementation of either wheat germ oil or corn oil. At the end of the trial, the researchers found improvements in markers of cardiovascular health in those who took wheat germ oil. For example, there was a reduction in oxidative stress. They concluded that "wheat germ oil is an important source of n-3 fatty acids, which may exert an antiatherosclerotic effect."[4]

USEFUL FOR HYDRATING SKIN

In a study published in 2010 in *Pharmacognosy Research*, researchers based in India examined the ability of six different herbal moisturizing products to hydrate the skin. They selected the moisturizers based on the "presence or absence of wheat germ oil and Aloe vera extracts." The cohort consisted of six groups of six healthy human volunteers; all of the volunteers were in their twenties. For three weeks, each of the groups tested one moisturizer, applying it twice each day to the forearm. In evaluating the actions of the different moisturizers, the researchers looked for "conductance, glow and appearance." The researchers found that all of the moisturizers enhanced the appearance of the skin. "Improvement in the appearance of skin supports the data for the increase in hydration." However, the moisturizer that contained both wheat germ oil and Aloe vera extract "produced the highest hydration effect." Moreover, "the formulations containing wheat germ oil and Aloe vera extract produced high skin hydration as compared to the formulations containing them separately."[5]

IMPROVES OVERALL HEALTH

In a study published in 2008 on the *Journal of the American College of Nutrition*, researchers from France investigated the ability of wheat germ and wheat germ oil, as a dietary source of vitamin E, to provide antioxidant protection to rats. The researchers began by dividing their rats into three groups of eight rats. The first group served as the control; the rats in this group were fed corn oil, which has low levels of vitamin E. The rats in the second group were fed wheat germ, and the rats in the third group were fed wheat germ oil. According to the researchers, the wheat germ and wheat germ oil diets contained the same amount of vitamin E. After three weeks, the rats were sacrificed. The researchers found that both the wheat germ and wheat germ oil-enhanced diets "significantly increased plasma and liver vitamin E levels." At the same time, the two diets "strongly decreased the susceptibility of heart and liver lipids to oxidation."[6]

OFFERS PROTECTION FROM EFFECTS OF RADIOTHERAPY

In a study published in 2011 in the *Journal of American Science*, researchers from Saudi Arabia and Egypt explained that radiotherapy is commonly used to treat human cancers. But, people are also exposed to low levels of radiation in everyday life—during some medical diagnostic procedures, air travel, or just spending time outside. Does wheat germ oil provide a degree of protection from radiation? The researchers tested this hypothesis in rats. They began by dividing their rats into six groups of ten rats. The rats in the first group were untreated; they were the controls. The rats in the second group were exposed to a single dose of radiation. The rats in the third and fifth groups were treated with wheat germ oil at 1 and 3ml/kg for three consecutive days. The rats in the fourth and sixth group were treated with wheat germs oil at the same dose and the same three days and then exposed to radiation. The researchers found that wheat germ oil did offer a degree of protection from the radiation. And, they concluded that "the treatment of rats with what germ oil either at 1ml or 3ml/kg body weight prior to whole body γ-irradiation seems to exert protective effects."[7]

MAY OFFER PROTECTION FROM A HIGHLY TOXIC INSECTICIDE

In a study published in 2011 in *Ecotoxicology and Environmental Safety*, researchers from Turkey examined the ability of wheat germ oil to provide a measure of protection to mice exposed to coumaphos, a highly toxic insecticide. The researchers began with 48 mice, which they placed in one of four groups. The mice in the first group became the controls; they were given corn oil. The mice in the second group received 1.5ml/kg bw/day of wheat germ oil. The mice in the third group were given 5.5mg/kg bw/day of coumaphos. And, the mice in the fourth group were given the previously noted amounts of both wheat germ oil and coumaphos. After 45 days, the mice were sacrificed. The researchers found that "coumaphos led to adverse alterations in the majority of the oxidative stress markers investigated." At the same time, "the administration of wheat germ oil alleviated the coumaphos-induced adverse effects detected in the tissues examined."[8]

MAY HAVE ANTIDIABETIC PROPERTIES

In a study published in 2010 in the *Indian Journal of Medical Sciences*, researchers from India wondered if a combination of wheat germ oil, coriander, and aloe vera would have antidiabetic activity in rats. The researchers began by dividing the rats into 12 groups. Some of the rats served as controls; some rats remained normal, while others were experimentally induced with diabetes. Some rats were treated with a diabetes medication known as glibenclamide; many rats were treated with different doses and ratios of the "polyherbal" product. The trial continued for 30 days. The researchers found that the polyherbal combination had a definite antidiabetic effect. It lowered glucose levels in normal and diabetic rats. They concluded that the "polyherbal preparation, a combination of three herbal plants, exerted a significant anti-diabetic effect."[9]

IS WHEAT GERM OIL BENEFICIAL?

For most people, wheat germ oil appears to be an excellent addition to the diet. However, people who have a gluten or wheat sensitivity or who have celiac disease must not use the most common forms of this oil. They contain wheat and must be avoided.

NOTES

1. Livestrong.com.
2. Hassanein, Minar Mahmoud M. and Adel Gabr Abedel-Razek. 2009. "Chromatographic Quantitation of Some Bioactive Minor Components in Oils of Wheat Germ and Grape Seeds Produced as By-Products." *Journal of Oleo Science* 58(5): 227–233.
3. Brandolimi, Andrea and Alyssa Hildago. March 2012. "What Germ: Not Only a By-Product." *International Journal of Food Sciences and Nutrition* 63 Supplement 1: 71–74.
4. Alessandri, Cesare, Pasquale Pignatelli, Lorenzo Loffredo et al. 2006. "Alpha-Linolenic Acid-Rich Wheat Germ Oil Decreases Oxidative Stress and CD40 Ligand in Patients with Mild Hypercholesterolemia." *Arteriosclerosis, Thrombosis, and Vascular Biology* 26: 2577–2578.
5. Saraf, S., S. Sahu, C. D. Kaur, and S. Saraf. May 2010. "Comparative Measurement of Hydration Effects of Herbal Moisturizers." *Pharmacognosy Research* 2(3): 146–151.
6. Leenhardt, Fanny, Anthony Fardet, Bernard Lyan et al. April 2008. "What Germ Supplementation of a Low Vitamin E. Diet in Rats Affords Effective Antioxidant Protection in Tissues." *Journal of the American College of Nutrition* 27(2): 222–228.
7. Barakat, Ibrahim A. H., Osama A. Abbas, Samia Ayad, and Aziza M. Hassan. 2011. "Evaluation of Radio Protective Effects of Wheat Germ Oil in Male Rats." *Journal of American Science* 7(2): 664–673.
8. Karabacak, M., M. Kanbur, G. Eraslan, and Z. Soyer Sanca. October 2011. "The Antioxidant Effect of Wheat Germ Oil on Subchronic Coumaphos Exposure in Mice." *Ecotoxicology and Environmental Safety* 74(7): 2119–2125.
9. Srivastava, Noopur, Gaurav Tiwari, and Ruchi Tiwari. April 2010. "Polyherbal Preparation for Anti-Diabetic Activity: A Screening Study." *Indian Journal of Medical Sciences* 64(4): 163–176.

REFERENCES AND RESOURCES
Magazines, Journals, and Newspapers

Alessandri, Cesare, Pasquale Pignatelli, Lorenzo Loffredo et al. 2006. "Alpha-Linolenic Acid-Rich Wheat Germ Oil Decreases Oxidative Stress and CD40 Ligand in Patients with Mild Hypercholesterolemia." *Arteriosclerosis, Thrombosis, and Vascular Biology* 26: 2577–2578.

Barakat, Ibrahim A. H., Osamaa A. Abbas, Samia Ayad, and Aziza M. Hassan. 2011. "Evaluation of Radio Protection Effects of What Germ Oil in Male Rats." *Journal of American Science* 7(2): 664–673.

Brandolini, Andrea and Alyssa Hildalgo. March 2012. "Wheat Germ: Not Only a By-Product." *International Journal of Food Sciences and Nutrition* 63 Supplement 1: 71–74.

Hassanein, Minar Mahmoud M. and Adel Gabr Abedel-Razek. 2009. "Chromatographic Quantitation of Some Bioactive Minor Components in Oils of Wheat Germ and Grape Seeds Produced as By-Products." *Journal of Oleo Science* 58(5): 227–233.

Karabacak, M., M. Kanbur, G. Eraslan, and Z. Soyer Sanca. October 2011. "The Antioxidant Effect of Wheat Germ Oil on Subchronic Coumaphos Exposure in Mice." *Ecotoxicology and Environmental Safety* 74(7): 2119–2125.

Leenhardt, Fanny, Anthony Fardet, Bernard Lyan et al. April 2008. "Wheat Germ Supplementation of a Low Vitamin E Diet in Rats Affords Effective Antioxidant Protection in Tissues." *Journal of the American College of Nutrition* 27(2): 222–228.

Saraf, S., S. Sahu, C. D. Kaur, and S. Saraf. May 2010. "Comparative Measurement of Hydration Effects of Herbal Moisturizers." *Pharmacognosy Research* 2(3): 146–151.

Srivastava, Noopur, Gaurav Tiwari, and Ruchi Tiwari. April 2010. "Polyherbal Preparation for Anti-Diabetic Activity: A Screening Study." *Indian Journal of Medical Sciences* 64(4): 163–176.

Website

LIVESTRONG.COM. http://livestrong.com.

Glossary

Aerodigestive The combined organs and tissues of the respiratory tract and the upper part of the digestive tract.

Alloxan An oxidized product of uric acid that destroys islet cells of the pancreas, thereby causing diabetes.

Androgenetic Alopecia The absence of hair from skin areas in which it is normally present. It is the most common type of hair loss. It is due to male hormones.

Angina A type of chest pain caused by reduced blood flow to the heart muscle.

Anthropometric Human body measurements.

Antiandrogenic A substance that blocks the actions of androgens, the hormones responsible for male characteristics.

Antiatherogenic Activity Helps to prevent the formation of plaque.

Antiproliferative Inhibit the growth of cells as in inhibiting the growth of cancer cells.

Antipyretic Has properties that reduce fever.

Apoptosis Cell death.

Atherogenesis Formation of fatty lesions on arterial walls.

Atherosclerosis Also known as hardening of the arteries. It is the buildup of fat, cholesterol, and other substances in the walls of arteries and the formation of hard structures called plaque.

Atopic Dermatitis A long-term skin disorder in which there are scaly and itchy rashes.

Atrial Fibrillation An erratic heart rhythm originating in the atrium.

Benign Prostatic Hyperplasia Enlargement of the prostate gland.

Body Mass Index (BMI) A body measurement based on a person's height and weight.

Cheilitis Painful inflammation and cracking of the lips.

Chondroitin   A molecule occurring naturally in the body that is a major component of cartilage.

Citraturia   Levels of citric acid in the urine.

Copra oil   Oil made from dried coconut flesh.

Coumaphos   A highly toxic insecticide.

C-Reactive Protein   A protein produced by the liver. It rises when there is inflammation in the body.

Crackles   Added breath sounds heard in the lungs of people with respiratory illness, such as pneumonia.

Cytotoxic   Toxic to living cells.

Dyslexia   A developmental reading disorder in which the brain has a specific information processing problem.

Dyslipidemia   An abnormal amount of lipids in the blood.

Dysmenorrhea   Pain during menstruation.

Dyspepsia   Indigestion.

Ellagic Acid   A phytochemical found in some plant foods such as strawberries and raspberries. It is believed to have anticancer properties.

Enterocolitis   Inflammation of the colon and small intestine.

Erythema   A skin condition characterized by redness or a rash.

Erythrocytes   Red blood cells.

EsophagoGastroDuodenoScopy   A diagnostic endoscopic procedure used to see the upper portion of the gastrointestinal tract.

Essential Fatty Acids   Fatty acids that the body requires to function, but the body is unable to produce them.

Fasting Plasma Glucose   Also known as the fasting blood sugar test. It measures blood sugar levels when the patient is fasting.

Granuloma   Small nodule.

Hemodynamic   Movement of blood.

Hepatotoxicity   Liver injury caused by drugs.

Hypercholesterolemia   Elevated levels of serum cholesterol.

Hyperlipidemia   High levels of lipids in the blood.

Hypertension   Elevated blood pressure.

Hypoxia   Deprived of adequate oxygen supply.

Inguinal Hernia   The protrusion of intestine in the groin area.

Insulin Resistance   An impaired ability of cells to respond to insulin.

Irritable Bowel Syndrome   A disorder in which there are recurrent episodes of abdominal distension and bloating, abdominal pain, and altered bowel habits with constipation, diarrhea, and an urgency to defecate.

Isoenergetic   Equal.

Isoflavones   Phytochemicals or compounds produced by plants that may have an effect on the body.

Legumes   A family of plants that bear edible seeds in pods.

Lipemia   Abnormally high concentrations of lipids in the blood.

Lipids   A broad group of naturally occurring molecules that include fat, waxes, sterols, and fat-soluble vitamins.

Liposomes   Microscopic artificial sacs composed of fatty substances that contain a water droplet mixture and an active ingredient. It is often used in experimental research.

Lycopene   A carotenoid that gives fruits and vegetables a red color.

Mastalgia   Breast pain.

Metabolic Syndrome   Also known as insulin resistance. With this disorder, there is extra weight around the waist, high blood pressure, and elevated levels of cholesterol.

Metastasis   The spreading of cancer from its initial site to other parts of the body.

Mucositis   Breaking down of mucous membranes leading to ulceration and infection.

Neoplasm   A new and abnormal growth, especially characteristic of cancer.

Nociception   Pain from the stimulus of nerve cells.

Nocturia   Getting up during the night to urinate.

Obesogenic   Supportive of obesity, promoting obesity.

Oleuropein   A phenolic compound found in olive oil.

Osteoarthritis   A degenerative joint disease characterized by damage to cartilage, which cushions the joints.

Osteocalcin   A marker of bone formation.

Osteopenia   Bone mineral density that is lower than normal but not low enough to be classified as osteoporosis.

Osteoporosis   A condition in which the bones lose too much mineralization. The weakened bones are at increased risk for breakage.

Ozone   A gas comprised of three oxygen atoms.

Parenteral   Introduction of nutrition, medicine, or other substances into body via a route other than the mouth.

Periodontal Disease   Inflammation of the gums and the loss of attachment of the periodontal ligament.

Photoaging   Damage to the skin from ultraviolet radiation.

Phytic Acid   A dietary fiber component found in most grains and legumes that has been shown to have antioxidant and anticancer properties.

Plant Polyphenols   Naturally occurring plant compounds that offer protection from a number of different illnesses.

Postprandial Lipemia   Elevated lipid levels after the consumption of a meal.

Proapoptotic   Programmed cell death.

Psoriasis   A common skin condition that causes redness and irritation. People with psoriasis often have thick red skin with flaky silver–white patches called scales.

Radical Prostatectomy   Removal of the prostate gland and surrounding tissue.

Reepithelialization   Restoring the soundness from injury of external surfaces of the body.

Rheumatoid Arthritis   A disorder in which the immune system attacks the joints, making them swollen, stiff, and painful.

Sarcopenia   The degenerative loss of skeletal muscle, muscle mass, and strength associated with aging.

Silver Sulphadiazine   A topical, antibacterial cream used for burns.

Sjogren's Syndrome   An autoimmune disorder in which the body attacks glands in the body that produce moisture, such as the salivary and tear glands.

Strangury   Frequent, painful urination of small amounts of urine.

Tear Film Osmolarity   The salt content of the eye, which tends to be higher in people with dry eyes.

Triglycerides   A type of fat in the blood that provide energy to the body.

Ulcerative Colitis   A type of inflammatory bowel disease that affects the lining of the large intestine and rectum.

Urolithiasis   Process of forming stones in the kidneys, bladder, or urethra.

Index

Acetylsalicylic acid (aspirin), 7, 140

Acne: coconut oil, 34–35; fish oil, 54; jojoba oil, 101–2; olive oil, 153

Acyclovir (antiherpes drug), 103

Aerodigestive tract, 120

Allergies: to jojoba oil, 103–4; to peanuts, 162; treating children with fish oil, 80–81; treating children with olive oil, 111–12; treating children with walnut oil, 214

Alloxan (cause of diabetes), 213

Almond oil, 1–6; anecdote to aluminum phosphide pesticide poisoning, 3–4; background, 1; benefits summary, 4–5; liver health, 2–3; preterm infants (topical massage), 3; prevention for striae gravidarum (stretch marks), 4; skin protection, 3; weight control and type 2 diabetes, 1–2

Aloe vera cream, for sulfur mustard (mustard gas), 152

Alpha-linoleic fatty acid (ALA), 186, 216, 217; See also Fish oil; Flaxseed oil; Krill oil; Omega-3 fatty acids; Salmon oil; Wheat germ oil

Aluminum phosphide pesticide: almond oil, 3–4; coconut oil, 30

Alzheimer's disease: coconut oil, 27; olive oil, 147–49; salmon oil, 191–93

Androgenetic alopecia (hair loss), saw palmetto oil, 196–97

Antiaging, safflower oil, 183

Antiatherogenic activity, sea buckthorn oil, 199

Antibacterial properties, sea buckthorn oil, 199–200

Anticoagulant properties, evening primrose oil, 50

Antidiabetic effects. See Diabetes; Metabolic syndrome; Type 1 diabetes; Type 2 diabetes

Antifungal properties, sea buckthorn oil, 199–200

Antihepatoxicity properties: almond oil, 1, 3; coconut oil, 33; See also Liver health/protection

Anti-hyperglycemic efficacy. See Metabolic syndrome; Type 2 diabetes; Type 1 diabetes; Diabetes

Anti-inflammatory properties: almond oil, 1; borage seed oil, 15; coconut oil, 29; fish oil, 54, 62–65, 71; flaxseed oil, 99; jojoba oil, 103; krill oil, 107; olive oil, 113, 143; walnut oil, 212

About the Authors

MYRNA CHANDLER GOLDSTEIN, MA, has been a freelance writer and independent scholar for 25 years. She is the author of *Healthy Foods: Fact versus Fiction* and *Healthy Herbs: Fact versus Fiction* and several other books with Greenwood Press, an imprint of ABC-CLIO.

MARK A. GOLDSTEIN, MD, is the founding chief of the Division of Adolescent and Young Adult Medicine at Massachusetts General Hospital. Dr. Goldstein, associate professor of pediatrics at Harvard Medical School, is author or editor of numerous professional and lay publications. His research interests include studying the effects of eating disorders and malnutrition on bone mineralization in adolescents and young adults.